Aesthetic Science

Aesthetic Science

Representing Nature in the Royal Society of London, 1650–1720

ALEXANDER WRAGGE-MORLEY

The University of Chicago Press
Chicago and London

The University of Chicago Press, Chicago 60637
The University of Chicago Press, Ltd., London

Published 2020
Printed in the United States of America

29 28 27 26 25 24 23 22 21 20 1 2 3 4 5

ISBN-13: 978-0-226-68072-9 (cloth)
ISBN-13: 978-0-226-68086-6 (paper)
ISBN-13: 978-0-226-68105-4 (e-book)
DOI: https://doi.org/10.7208/chicago/9780226681054.001.0001

Library of Congress Cataloging-in-Publication Data

Names: Wragge-Morley, Alexander, author.
Title: Aesthetic science : representing nature in the Royal Society of London, 1650–1720 / Alexander Wragge-Morley.
Description: Chicago ; London : University of Chicago Press, 2020. | Includes bibliographical references and index.
Identifiers: LCCN 2019047754 | ISBN 9780226680729 (cloth) | ISBN 9780226680866 (paperback) | ISBN 9780226681054 (ebook)
Subjects: LCSH: Ray, John, 1627–1705. | Boyle, Robert, 1627–1691. | Grew, Nehemiah, 1641–1712. | Hooke, Robert, 1635–1703. | Willis, Thomas, 1621–1675. | Royal Society (Great Britain) | Science—Great Britain—History—17th century. | Science—Aesthetics. | Knowledge, Theory of. | Senses and sensation—Great Britain.
Classification: LCC Q127.G4 W73 2020 | DDC 509.2/241—dc23
LC record available at https://lccn.loc.gov/2019047754

⊗ This paper meets the requirements of ANSI/NISO Z39.48-1992 (Permanence of Paper).

Contents

Introduction

> Since things themselves cannot be painted and heard, we paint and listen to representations of them. Even if these representations are not similar to them, we see nonetheless certain sensible beautiful things in them that make us understand a theorem, that is, a property of the intelligible thing itself.
>
> GOTTFRIED LEIBNIZ, "De Characteribus et Compendiis," 1676[1]

Given that the senses seem to provide us our only means of reckoning with the external world, questions about the meanings of sensory experience have long been of fundamental importance to the theory and practice of the sciences. Such questions became particularly urgent, however, during the second half of the seventeenth century, when many of the natural philosophers associated with the nascent Royal Society of London urged that sensory experience should serve as the foundation for a new science of nature. Indeed, this moment is usually regarded as one of the crucial phases in the emergence of the modern empirical sciences in Europe. At the time of the society's foundation in 1660, the suggestion that sensory experience could serve as the basis for a system of knowledge about the workings of the physical world was not as new as has often been claimed. Nevertheless, it is fair to say that the schools of thought then dominant—whether the broadly Aristotelian natural philosophy still practiced in many universities or the so-called new philosophy identified with René Descartes (1596–1650)—generally regarded experience as an unpromising foundation for knowledge, noting the many errors and uncertainties intrinsic to our sensory encounters with the world. For Descartes, the way to resolve those uncertainties was by turning away from experience and relying instead on the dead certainties of deductive reasoning. He and the other thinkers known today as rationalists, including figures such as Thomas Hobbes (1588–1679) and Baruch Spinoza (1632–1677), hoped to discover the truth by subjecting the world of experience to systems of thought derived from mathematics and logic.

Among the members of the Royal Society, however, philosophers such as Robert Boyle (1627–1691), Robert Hooke (1635–1703), and John Ray (1627–1705) held quite another position. Building on both ancient and modern forms of empiricism, they proposed that the most effective way to understand the world was by gradually building up a picture of its workings from observations and experiments—that is, by a process of induction in which the collection of many observations would eventually make it possible to grasp some of nature's general principles. Seeking to vindicate the Royal Society in the face of widespread criticism, the controversialist Joseph Glanvill (1636–1680) thus asserted that philosophers could obtain a more illuminating picture of nature "from the *Observations* and *Applications* of *Sense*" than through any kind of deductive reasoning.[2] It did not follow, however, that the proponents of this empirical approach had any more faith in the senses than Descartes did. On the contrary, Glanvill and others reflected at great length on the many ways in which sensory experience could prove misleading or even prejudicial to the work of natural philosophy. Perhaps the most fundamental of their doubts sprang from the observation that the senses did not appear to have been designed for the sake of knowledge production at all. As Hooke ruefully put it, the "Design [. . .] of the Organs of Sense, seems to have been for some other Use than for the acquiring [. . .] of Knowledge, and to have a very great Affinity with the Senses of other Animals." Far from serving the purposes of natural philosophy, Hooke suggested, the primary function of sensation was the preservation of animal life. The senses existed not to fill the mind with knowledge but rather to guide irrational animals—including humans in their animal aspect—away from things likely to do them harm, and toward those that might do them good.[3]

With this reflection on what we might call the animality of the senses, Hooke gestured toward one of the most troubling paradoxes of the empirical philosophy. For its information about the external world, the mind depended on sensory organs that did not always respond objectively to the things around them. In fields ranging from political thought to neurophysiology, thinkers of the seventeenth century recognized that those seemingly irrational responses could exercise considerable power over both the body and the mind. In *An Essay Concerning Human Understanding* (1690)—perhaps the single most influential attempt to describe the epistemology of empiricism—John Locke (1632–1704) argued that the feelings of pleasure arising from sensory experience motivated much human thought and action.[4] On the one hand, he regarded those pleasures as intrinsic to the exercise of reason, arising from fundamental cognitive operations such as the acquisition of ideas about the external world and the comparison of those ideas by the mind. On the other hand, Locke worried that the same feelings of pleasure might make the exercise of reason impossible. Indeed, he held that the body's arbitrary manner of annexing enjoyment to certain forms of experience was one of the leading causes of error, impelling individuals to choose what gave them pleasure over the less obviously gratifying work of seeking the truth.[5] For Locke and his contemporaries, the exercise of reason depended on the careful management of feelings that appeared to spring without any thought or reflection from sensory encounters with external things.[6]

For a long time, historians and philosophers of science argued that the Royal Society's solution to this problem was to find ways of disciplining affect out of experience, using philosophical method to anesthetize both individuals and the philosophical community against the distortions of individual subjectivity. They have suggested that the society developed practices that gave rise to a recognizably modern form of empiricism, characterized both by a dispassionate account of scientific experience and by skepticism about the possibility of using natural philosophy to make metaphysical claims, or other arguments about the ultimate reasons for things.[7] Developments in recent years, however, have begun to point out the inadequacy of that narrative. Historians such as Lorraine Daston, Jessica Riskin, Charles T. Wolfe, and Ofer Gal have all pointed out that scientists of the early modern period and beyond recognized the epistemic potential of seemingly subjective states, from the passions of desire to the kinds of instant recognition arising from perceptual habit. They have raised the possibility, in other words, that our image of the history of the empirical sciences requires reconsideration. Instead of telling a story about the ways in which scientists subjected the objects of experience to disembodied reason or some other form of affectless discipline, they have suggested that scientists found ways to incorporate their affective states and other subjectivities into the practices of scientific inquiry and representation.[8]

The explicitly empirical science promoted by the Royal Society in the seventeenth century should be an important target for just such a reconsideration. The image of the society as a body of philosophers committed to something that we would recognize as scientific objectivity, premised on a desire to produce transparent representations of the reality of the world and nothing more, remains surprisingly tenacious. Such a view, indeed, continues to inform assessments of the relationships between theology, metaphysics, and representational practices in the work of philosophers such as Robert Boyle, Nehemiah Grew, Robert Hooke, John Ray, and Thomas Willis. In recent years, however, focusing on the study of visual, material, and textual practices, scholars have begun to show that the provocation of feeling, not just the dispassionate exercise of reason, was central to the Royal Society's work. It is high time, therefore, to seek out ways of incorporating the history of experience—including its affective and subjective dimensions—into the history of the empirical sciences. In *Aesthetic Science*, I argue that we can only accomplish this end by making experience into our main category of analysis. The history of the empirical sciences must, in other words, be a history of experience itself.

The "Aesthetic" and the Empirical Sciences

The concepts commonly associated with aesthetics are powerful tools for working out how philosophers such as Hooke and Locke attempted to make the pleasures and pains of the senses serve the interests of knowledge production. However, the use of terminology alien not only to the time and place under consideration but also to the topic as it is conventionally understood requires explanation. After all, the term *aesthetic* is today inextricably bound up with an account of art—or rather, the forms of experience to be expected from it—that emerged over the course of the eighteenth century, most famously in the work of Immanuel Kant (1724–1804). In his *Critique of the Power of Judgment* (1790), Kant sought to show that judgments of beauty depended on mental processes distinct from those at stake in the exercise of rational thought. The special form of pleasure arising from the encounter with beautiful objects, he argued, was incompatible with the cognitive activity at stake in the determination of concepts. For Kant, aesthetic experience depended on a form of mental freedom—what he termed the "free play" of the mental faculties. There was simply no way for the mind to exercise this freedom, he argued, when it was forced to follow the logical procedures required for working out moral or scientific truths.[9]

Insisting on its irrational and apparently subjective qualities, Kant described a form of aesthetic experience that in some respects resembles the category of bodily and mental states known to philosophers and literary critics as *affects*. Including but not limited to feelings such as pleasure and pain, *affective* states are usually defined as states that arise in the mind or body prior to or outside rational thought, provoking responses without cognition or perhaps even consciousness.[10] In his *Rhetoric*, Aristotle (384–322 BCE) defined anger and other emotions as "affections" that could provoke people into changing their opinions without engaging in a process of reasoning.[11] In like fashion, the Dutch philosopher Baruch Spinoza portrayed the affects as forces that acted upon the mind even though the mind was incapable of reducing them to clear concepts. He thus famously defined *affectus* as "a confused idea [. . .] which being given, the mind itself is determined to thinking this rather than that."[12] There was a crucial difference, however, between

Kant's model of aesthetic experience and the affective states outlined by such thinkers as Aristotle and Spinoza. Despite arguing that the experience of beauty arose from an irrational—and thus subjective—form of mental activity, Kant sought to show that it had a claim to universality. Giving it many of the same qualities as an affect, he nevertheless insisted that the perception of beauty could command a form of universal assent, much like the sort that might be expected from the correct application of reason to scientific problems.[13]

Earlier in the eighteenth century, thinkers distinguished less clearly between the experience of sensory pleasure and the exercise of reason. In part, this difference arose because the field of aesthetics was not yet restricted to questions about beauty. Indeed, Alexander Gottlieb Baumgarten (1714–1762) introduced the term *aesthetic* into the European philosophical lexicon to name a field of inquiry that had a much wider scope. Defined by Baumgarten as the "science of sensitive cognition," that field was concerned not merely with the experience of art but more broadly with sensory perception and its links with the other faculties of the mind.[14] Philosophers produced different solutions, moreover, to questions about the relationship between the affective dimensions of experience and the exercise of reason. The Irish-Scottish philosopher Francis Hutcheson (1694–1746) recognized that aesthetic pleasure did not appear to depend on rational thought or even conscious reflection. In his *Inquiry into the Original of Our Ideas of Beauty and Virtue* (1725), he characterized beauty as a kind of internal sensation, arising instantly and without reflection from encounters with certain objects. Unlike Kant, however, Hutcheson did not separate the experience of that kind of pleasure from the exercise of reason. On the contrary, he asserted that scientific arguments—"Theorems, or universal Truths demonstrated"—gave rise to precisely the same kind of gratification that he associated with the beauties of nature and art. Hutcheson suggested, moreover, that God had arranged the otherwise inexplicable correspondence between such truths and the experience of aesthetic pleasure to encourage human curiosity.[15]

The aim of this book is not to adjudicate between these competing claims about the relationship between the experience of certain forms of pleasure and the exercise of reason. Rather, I propose to use aesthetic theory as a resource of concepts and questions—that is, as a set of tools with which to understand how an even earlier generation of natural philosophers reconciled the affective dimensions of experience, arising prior to or independent from cognition, with the desire to rationally obtain knowledge by empirical means. Arguing that some of the leading members of the seventeenth-century Royal

Society pursued an "aesthetic science," I do not mean to suggest that they identified the discovery of the truth with the experience of beauty as it came to be defined by Kant. Nor do I simply repeat the now well-established observation that, in the early modern period, there were close connections, both conceptual and practical, between the arts and the sciences. My aim instead is to show that concerns about the affective dimensions of sensory perception came, in crucial respects, to define their approaches to the production of knowledge. As we shall see, those concerns spanned the whole range of issues addressed in later works of aesthetic theory, from questions about the relationships between sensory perception and the exercise of reason to the problem of taste—the difficulty of getting people to agree about apparently subjective experiences such as beauty and pleasure. Moreover, the resources of aesthetics provide us with a means of revealing hitherto overlooked connections between the metaphysics, theology, and scientific practices of early Royal Society members. Focusing on their preoccupation with the affective dimensions of experience, I will therefore also seek to shed new light on the role of metaphysical and theological ideas in their visual, material, and textual practices.

We can get an initial sense of the role that aesthetic problems had in the empirical project by considering one of the first works published under the Royal Society's imprimatur—Robert Hooke's *Micrographia; or, Some Physiological Descriptions of Minute Bodies*, released in 1665. One of the earliest books to include images of natural things viewed under optical magnification, the *Micrographia* has played a significant role in attempts to explain the society's collective approach to the production of knowledge.[16] In the preface, Hooke made the case for a form of knowledge production that conforms to the modern image of the empirical sciences, arguing that the main task of natural philosophy was the production of faithful representations of the world as it presented itself to the senses. He also filled the book with beautiful images of the phenomena that he had investigated. Like the engraving of a flea that we see in figure 1, those images seem far too polished to be mere transcriptions of Hooke's experience.

One explanation is that Hooke included such images for rhetorical purposes, seeking to promote natural philosophy by making its objects seem far more beautiful than they really were. Since the late twentieth century, however, scholars have offered altogether more sophisticated accounts of Hooke's visual rhetoric, arguing that he sought to produce a palpable—albeit highly artificial—effect of direct observation. Meghan C. Doherty has shown, for instance, that Hooke used his familiarity with the techniques and codes of

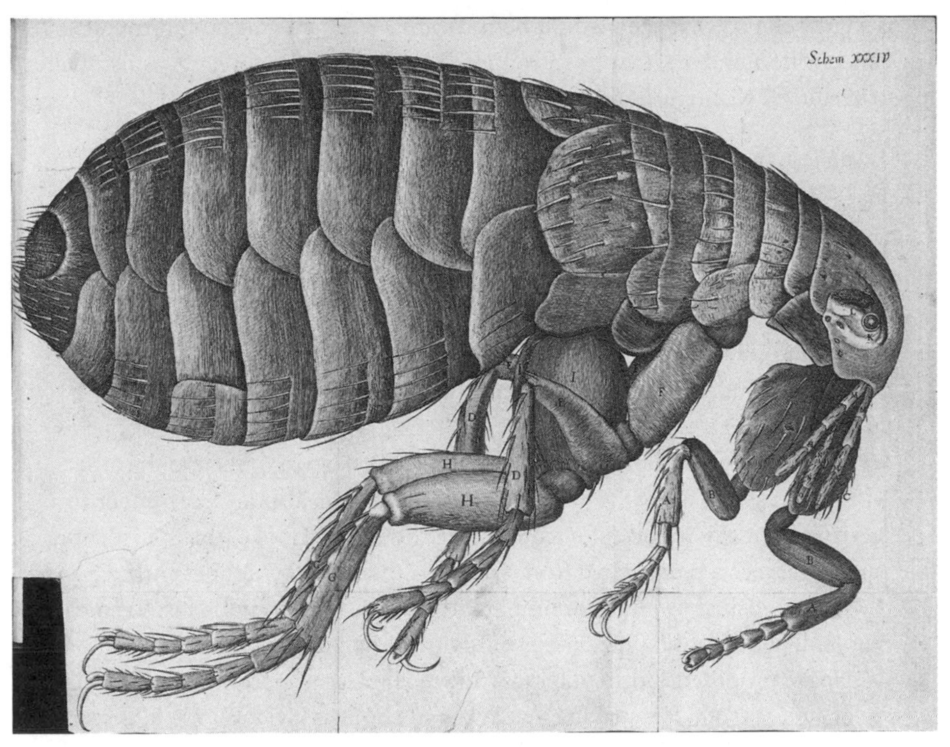

FIGURE 1. Fold-out engraving depicting a flea under microscopic observation. Robert Hooke, *Micrographia* (1665), Scheme XXXIV. Wellcome Library, London. Creative Commons CC BY.

portrait engraving to "construct an illusion of the act of looking at the visible world." Rather than simply reproducing the data of experience in pictorial form, he mobilized contemporary artistic practices to give his readers the feeling that they were participating in his observational acts. Doherty therefore reminds us not only that the production of scientific images depended on artistic expertise but that scientific objectivity itself was a cultural construct, dependent on changing visual practices and conventions.[17]

Hooke also used his images, however, to substantiate an argument about the experience of natural philosophy. He asserted that the study of nature should provoke pleasure, repeatedly insisting that the mineral, plant, and animal bodies he had placed under magnification were "beautiful," "curious," or "exceeding pleasant."[18] For Hooke, those pleasures were not just subjective experiences. Instead, he regarded them as necessary outcomes of the relationship between God and the world that God had created. The belief that God was an infinitely wise, powerful, and benevolent creator led Hooke to expect not only that natural things would exhibit a high degree of perfection in their design but also that they would be beautiful. Such was his commitment to this position, indeed, that he introduced the experience of beauty as his sole criterion for distinguishing between natural and artificial things:

> So unaccurate is [Art], in all its productions, even in those which seem most neat, that if examin'd with an organ more acute than that by which they were made, the more we see of their *shape*, the less appearance will there be of their *beauty*: whereas in the works of *Nature*, the deepest Discoveries shew us the greatest Excellencies.

Hooke's point was that even the finest pieces of human workmanship appeared ugly under the magnifying lens, while the products of God's design, by contrast, became even more attractive. Hooke positioned the microscope as something more than an instrument for yielding deep insights into the otherwise inaccessible workings of natural things. It was also an instrument for rectifying the experience of encountering them. For Hooke, the microscope gave observers a perspective from which to discover the ugliness lurking in the most polished products of human art and the beauty hidden in the most outwardly unappealing specimens of divine design—like his lovely flea.[19]

Hooke proposed a form of inquiry in which pleasure accompanied the discovery of the truth. His striking images served both as vivid representations of the things that he had encountered and as attempts to communicate the gratification that he claimed to have experienced. Indeed, Hooke recognized that not everyone felt the pleasure he expected from the products of divine design, attributing that failure to the corruption of the human faculties following Adam's ejection from the Garden of Eden. Using the microscope, he suggested, individuals could reverse the effects of sin on sensory perception, regaining a forgotten capacity to experience the necessary perfection and beauty of God's creation.[20] The affective dimensions of experience thus linked Hooke's material practices to a metaphysics of divine design. Shifting our attention to broad questions about sensory experience, we can therefore do more than point out once again that natural philosophers made use of contemporary artistic techniques and conventions in the communication of knowledge. In the following chapters, we shall see that concerns about the affective dimensions of experience animated a wide range of intellectual and material practices, including metaphysics, theology, the neurosciences, and the techniques of representation. Gathering these concerns into the broad category of aesthetic experience, rather than examining them through modern intellectual categories, will make it possible to bring the wide range of disciplines at stake in early modern natural philosophy into a single focal point. In so doing, we shall find that apparently subjective feelings of beauty and pleasure—often seen both then and now as prejudicial to the discovery of the truth—were far more important to the empirical project than has yet been understood.

Design, in Theory and in Practice

Robert Hooke's ideas about how nature should be experienced were closely related to his understanding of God's relationship to the world. Like many others in the Royal Society and beyond, drawing on an ancient tradition of using the study of nature for theological purposes, he argued that empirical natural philosophy could yield powerful evidence for both God's existence and God's close involvement with creation. Until the end of the seventeenth century, the dominant form of this argument, broadly speaking, was the one presented in the *Micrographia*. Along with members of the Royal Society such as Robert Boyle, Nehemiah Grew (ca. 1641–1712), John Ray and Thomas Willis (1621–1675), he claimed that the most compelling rational argument for God's involvement with the world was the design supposedly to be found in natural things. Based on the discovery of purposive mechanisms in the bodies of plants and animals, this version of the design argument went on to exercise immense influence, serving throughout the eighteenth century as one of the two leading scientific arguments for the existence of God. As is well known, the second of those was another version of the design argument, often practiced alongside the first. This second way of reasoning about the existence and attributes of God was based not on close encounters with the bodies of plants and animals but rather on the orderly cosmos described by Isaac Newton (1642–1727) and his followers from the turn of the eighteenth century onward.[21]

The significance of these arguments for our understanding of what it meant to use the senses in the production of knowledge has been to a large degree obscured, however, by a propensity to regard them as expressions of religious belief, little relevant to what natural philosophers did when they investigated and represented things. In the middle decades of the twentieth century, historians such as Richard Westfall generally regarded the design argument as the vestigial trace of an older religious mentality, sitting uneasily alongside an otherwise recognizably modern form of empiricism. Few scholars would speak in such a manner today. In recent years, historians such as William Poole have instead shown that practices such as the interpretation of sacred texts were central to problems in natural philosophy, especially when the Bible could be regarded as a pertinent source of evidence for natural processes. Such cases included debates about the meaning of fossils and the age of the earth, along with attempts to purify language to make it a more effective vehicle for philosophy—involving speculation about the language spoken by Adam in the Garden of Eden.[22] Nevertheless, the older distinction between what Peter Harrison calls the "territories" of religious belief and scientific knowledge continues to exert a subtle influence. Historians working on the interconnections between science and religion, for instance, still generally regard the design argument as little more than an apologetic strategy, used by naturalists of the seventeenth century to provide a religious justification for the empirical study of nature. As a result, they have done little to explore what role the idea of a world designed by God may have played in the work of natural philosophy.[23] In the scholarship concerned with the visual and material culture of the Royal Society, meanwhile, scant attention has so far been paid to religion. Despite the growing awareness that natural philosophy was invariably tied up with theological questions and the practices used to resolve them, the role of claims about design in the society's material and visual practices remains largely unknown.

There can be no doubt that the design argument reflected the Royal Society's corporate claims about the moral and political utility of natural philosophy. Founded in the year of Charles II's restoration to the English throne, the society positioned itself as a body that could assist in the reconstitution of a social order that, at least in the most widespread view at the time, had been shattered by unprovable claims to authority in matters of religion. Spokesmen such as Thomas Sprat (ca. 1635–1713) held up the Royal Society's brand of natural philosophy as a form of knowledge production that could put to rest heated religious debates, instead generating theological knowledge to which any reasonable person could consent.[24] At the same time, leading members of the Royal Society wanted to fend off the accusation that the new natural philosophy and its mechanical models of causation tended toward irreligion, whether by excluding God from the day-to-day workings of the world or by diverting attention toward worldly matters instead of those of the spirit. As William Poole and Rhodri Lewis have both pointed out, those criticisms were sometimes well founded. Even among the members of the society, there were those—such as Francis Lodwick (ca. 1619–1694) and William Petty (1623–1687)—who either reached heterodox conclusions when they applied natural philosophy to theology or doubted that natural philosophy could yield useful insights into the being and attributes of God.[25] In response, pious philosophers such as Boyle and Ray held up the design argument as the ultimate proof that natural philosophy could serve the interests of both conventional religious belief and a social order premised on the exercise of reason rather than the enthusiastic passions of religious inspiration.

It would be a mistake, however, to assume that those natural philosophers who supported the design argument had a single, clearly defined political agenda. Despite their show of support for the Restoration political order, indeed, only a handful of the philosophers whom we shall encounter had a straightforward allegiance to the Church of England. The botanist and theologian John Ray, for instance, had to give up his position at the University of Cambridge in 1662 because he felt unable to renounce the oaths he had sworn to the former regime.[26] Nehemiah Grew, Ray's counterpart in botany, came from a family of dissenters and never reconciled himself to the officially approved forms of religion.[27] Conformity, meanwhile, did not necessarily go together with politically congenial expressions of theological orthodoxy. The physician Thomas Willis, a friend of the new establishment, used the anatomical investigation of the brain and nerves to show that humans were unique in the possession of an immortal soul. Yet his proposal that the mind's simpler operations depended not on that soul but rather on mechanical agents common to humans and animals alike raised discomfiting questions about the extent of God's involvement with the day-to-day workings of the world.[28] If the proponents of the design argument were united only by loose political and theological allegiances, we will nevertheless see that they had one crucial thing in common. They all held that—through the discovery of design in nature—the senses could supply the mind with knowledge of not only the physical world but also the immaterial world of God. Like Hooke, moreover, they all asserted one way or another that the encounter with specimens of that design should be accompanied by pleasure. Hooke and his contemporaries proposed a natural philosophy that aimed at the discovery of design in natural things, identifying the experience of pleasure as a symptom of the encounter with a perfection that could only have come from God.

Developments in the historical study of religion may encourage us to reconsider the interconnections between religious ideas and scientific practices. Rather than regarding religion as a body of doctrines to be either defended or attacked, historians of religion have grown increasingly interested in seeing religion as a set of practices. Instead of attempting to measure belief in terms of adherence to a set of religious doctrines, in other words, they examine how individuals and groups participated in practices connected to religion, whether rituals, pedagogy, or even (for instance) the destruction of monasteries.[29] Adopting a similar perspective on the importance of practices rather than beliefs, historians of science now construct much broader accounts of the interconnections between science and religion, finding links between religious and scientific practices even when the individuals concerned did not necessarily share the same beliefs. In his book *The Good Life in the Scientific Revolution* (2006), for instance, Matthew L. Jones shows that the mathematics of René Descartes, Blaise Pascal, and Gottfried Leibniz owed much to a form of spiritual meditation that emerged during the sixteenth century, exemplified by the *Spiritual Exercises* of Ignatius of Loyola (1491–1556). Jones shows that philosophers of the seventeenth century saw mathematics as a form of spiritual cultivation—a practice that would make the mind fit for the pursuit of a good life.[30] Working from entirely different disciplinary perspectives, Sorana Corneanu and Joanna Picciotto have each demonstrated that thinkers such as Boyle and Locke also regarded the study of nature as a kind of moral praxis, serving the interconnected ends of uncovering the truth and preparing the philosopher for a life of virtue.[31] Historians of art and material culture, meanwhile, have sought to relate the visual practices of religion to broader transformations in visual and material culture. In a recent edited volume, to take just one example, Jeffrey Chipps Smith uses the expression "visual acuity" to describe an early modern visual praxis that cuts across today's disciplinary categories, ranging from the special kind of attentiveness required for the contemplation of sacred images to the focused observation necessary for obtaining knowledge.[32]

There is much to be gained by thinking about the design argument in the same manner. In the early Royal Society, after all, design was not just a theory—it was a practice. Few readers of this book will need to be reminded that Hooke and his counterpart Christopher Wren (1632–1723) devoted much time to designing buildings, justly remaining famous to this day for their contributions to the reconstruction of London after the Great Fire of 1666.[33] However, it is perhaps less well known that architectural design was also, as Kelsey Jackson Williams and others have shown, central to the antiquarian and historical interests of members including John Aubrey (1626–1697), Hooke, and Wren. Indeed, the material and intellectual practices of design—by no means limited to architecture—furnished members of the Royal Society with tools for recovering lost buildings such as Stonehenge and the Temple of Solomon, whether from surviving ruins or from ancient testimonies.[34] In this book, we will see that similar kinds of activity were crucial to the work of natural philosophy. From practices such as architecture, Hooke and his contemporaries took strategies for interpreting a world that they believed to be filled with the works of a divine designer. Those strategies included graphic and verbal practices—such as the techniques of architectural drawing—for uncovering the intentions of a designer from the otherwise undifferentiated field of human perception and for communicating those intentions forcefully to others. The practices of design were loaded, moreover, with what we might call "aesthetic expectations." Whether engaged in the study of artificial or natural things, philosophers such as Boyle, Grew, Hooke, and Ray expected the works of a good designer to exhibit high standards of beauty and perfection. Design therefore gave them more than a theoretical explanation—based on God's relationship to creation—for the apparent purposiveness on display in nature. In many instances, it gave them resources for transforming the unsynthesized manifold of sensory experience into empirical evidence for design.

It should now be clear that the point of taking an aesthetic approach to the design argument is not to ignore the place of religion in the brand of natural philosophy promoted by many members of the Royal Society. Rather, the point is to show just how important the ontological and aesthetic presuppositions of design were to the practices with which some of the society's leading members inquired into and represented the world. As the book unfolds, this demonstration will in turn make it possible to question the still-prevalent view that the Royal Society pioneered a recognizably modern form of empiricism, seeking simply to represent natural phenomena as they presented themselves to the senses. Of course, few scholars today would express this view so bluntly. Nevertheless, the continuing tendency to regard the design argument as an artifact of theological debate, little relevant to the practices of natural philosophy, leaves two of its essential components intact. First, it reinforces the notion that Hooke and his contemporaries practiced a form of empiricism that had little to do with the ontological assumption that nature was a product of divine design. Second, it gives us the false impression that the aesthetic corollary of that assumption—that the encounter with design would induce an experience of pleasure—had no role in the forms of experience at stake in natural philosophy. Recognizing the place of design in the practices of empiricism, by contrast, entails an acknowledgment that some of the leading natural philosophers of the seventeenth century saw the study of nature as a pleasurable encounter with the wisdom of God. For Boyle and his contemporaries, the senses could serve as a conduit for knowledge not only of the physical world open to observation but also of the world of imperceptible entities inhabited by God. In crucial respects, moreover, they practiced a form of empiricism that depended on the pleasures of sensory perception. Pleasure, as they saw it, was a symptom of the sensory encounter with specimens of God's wisdom and power.

Texts, Images, and Aesthetic Experience

Reconstructing the affective states that early modern philosophers attached to the experience of design in nature, however, is no straightforward task. Today's onlookers appear capable of sharing the feelings of pleasure and wonder that Hooke sought to evoke with the powerful illustrations that he included in the *Micrographia*. One might wonder, however, exactly what feelings he might have associated with the less obviously appealing images making up the bulk of the Royal Society's visual archive, whether an architectural drawing depicting only the floor plan of an ancient Roman temple, or a section taken through the root of a vine, delineating only the geometries of its passages and pores. Matthew Hunter has shown that the production and consumption of such visual materials was central to the work of natural philosophy. Experimentation with representational techniques played a generative role in the formation of hypotheses, even offering models for the cognitive operations at play in the acquisition of ideas through sensory experience.[35] To fully grasp the role of the society's visual archive in the production of knowledge, however, we must reckon with the fact that, as we shall see in greater detail, the pleasures as well as the pains of experience were widely thought to play a crucial role in those cognitive operations. The epistemic value of images was bound up with affective meanings that are by no means obvious today.

The difficulty of reconstructing the affective regime of empiricism seems even greater if we turn from visual to textual materials. With a few notable exceptions, the members of the Royal Society filled even their most heavily illustrated works with long, detailed descriptions. In many cases, moreover, they had to rely largely or even exclusively on verbal representations, whether because the costs of including images were too high or, as we shall see, because they worried that pictorial representations could prove misleading. For the most part, those verbal representations strike today's readers as matter-of-fact descriptions, with no obvious affective resonance other than dispassionate neutrality. Let us consider just one example. Depicting the fruit taken from a doum palm (here called a "date-nut"), it comes from Grew's catalog of the society's collection of natural and artificial rarities, *Musæum Regalis Societatis* (1681):

> The DATE-NUT, *qu. Nucidactylus*. 'Tis above two inches long; near the Stalk, above an inch over; towards the top near two, being belly'd like a *Pear*. Along one side, a little ridged. The Stalk cover'd with a whitish *Down*, like a *Quince's*. The outward Skin of a dusky Bay, smooth, soft, and thin. Next under this a *Work* of *Fibers*, not produced, as in other Fruits, by the length, but standing bolt upright, like the *Pile* of *Velvet*, about ¼ of an inch in depth; or rather, like the *Bristles* upon a *Hogs* back. So that the outward Skin being taken off, the Fruit looks and feels like a round *Scrubing-Brush*.[36]

Grew and his contemporaries presented descriptions like this as examples of a plain style that would promote clear and unambiguous communication. Packing their descriptions with comparisons to familiar objects such as scrubbing brushes rather than allusions to ancient mythology, they purported to reject older styles of writing that privileged the pleasures of rhetoric over the transmission of the truth. Indeed, they sometimes claimed to reject entirely the affective power long associated with poetry and rhetoric, asserting that their comparatively plain style served no other purpose than the transposition of reality into verbal form. Little wonder, therefore, that scholars working in the middle decades of the twentieth century, such as Richard Foster Jones, took those claims at face value, interpreting the Royal Society's descriptive style as an expression of the emergence of a modern scientific mentality.[37]

A growing awareness of the place of rhetoric and poetics in early modern theories of cognition has set the stage, however, for this view to be reconsidered. Peter Galison, Stephen Gaukroger, and Matthew L. Jones have all shown, for instance, that Descartes drew on an ideal of the vivid and pleasurable mental picturing long associated with powerful oratory when he first framed the argument that knowledge could only result from the possession of "clear and distinct" mental ideas. Despite claiming that the certainties of deduction provided the only sure path to the truth, Descartes modeled his account of knowledge on an apparently subjective description of the mental clarity arising from persuasive acts of speech.[38] A similar attentiveness to the influence of poetry and rhetoric, meanwhile, has led to a complete reinterpretation of the Royal Society's ambitious programs for linguistic reform. The changing fortunes of John Wilkins's famous proposal for an artificial language, published in 1668 as *An Essay towards a Real Character, and a Philosophical Language*, neatly exemplify this transformation. Generations of critics once characterized Wilkins's work as a fruitless attempt to make language into a completely transparent system of signs standing isomorphically for real objects of experience, reflecting a wider desire among members of the society to somehow eliminate the destabilizing pleasures of metaphor from philosophical discourse. William T. Lynch, Peter Walmsley, and Courtney Weiss Smith have demonstrated, however, that the Royal Society's critique of rhetoric concealed a deeper desire to make the pleasures of figurative speech serve the epistemologically virtuous purposes of natural philosophy. Indeed, Smith has shown that Wilkins built a scheme of rhymes and metaphors into his supposedly affectless language. Making use of the well-known cognitive value attached to rhyme and metaphor in theories of poetry and rhetoric, he sought a language where the relationships between different objects would be vividly expressed in the forms and sounds of the words.[39]

Recent work on the links between rhetoric and cognition in early modern philosophy therefore makes it possible for us to reconsider the descriptive style exemplified by Grew's lucid depiction of the fruit of the doum palm.

This reconsideration will take us beyond the now familiar observation that members of the Royal Society used rhetoric to persuade readers that what they reported was true.[40] Instead, it will hinge on a recognition that the affective states described in rhetoric and poetics had a significance that went far beyond their ostensible function. Until the middle of the eighteenth century, as we shall see, rhetoric was one of the most important domains for describing and theorizing, under the early modern rubric of the passions, the affective parts of human psychology.[41] Indeed, rhetoric was tightly interwoven into the fabric of virtually every discourse concerned with the provocation of passionate responses in the mind and body, ranging from art theory, in most of the forms in which it then existed, to treatises on the passions, pedagogy, politics, and preaching. Rhetoric was profoundly marked, moreover, by new attempts to explain the workings of the brain and the senses, most notably works of neurophysiology such as Descartes's *L'homme* (*Treatise on Man*, posthumously published in 1662) and Thomas Willis's anatomy of the brain, including its description and uses, *Cerebri Anatome, cui accessit Nervorum Descriptio et Usus* (1664). In his influential treatise on the arts of speech, *De l'art de parler* (1675), for instance, the French mathematician Bernard Lamy (1640–1715) used Descartes's physiology of perception and imagination to explain the persuasive force and cognitive effects of figurative language. Rhetoric stood at the heart of a nexus of discourses and practices concerned not only with the affective life of the mind but also with the very mechanics of sensory experience and cognition.[42]

Our inquiry into the rhetoric and poetics of verbal descriptions will thus facilitate a broader reconsideration of what it meant in the seventeenth century to obtain knowledge through the senses. This reconsideration has methodological implications that are worth exploring from the outset. For the most part, contributions to the burgeoning field concerned with the study of scientific images mobilize the analytical tools and categories belonging either to art history or to visual studies, privileging visual and material things at the expense of other forms of evidence.[43] With a few exceptions, meanwhile, the study of rhetoric and poetics in scientific writing remains the preserve of those who work on what we would today recognize as texts, such as scholars dealing with the history of literature, broadly conceived.[44] The importance of theories of verbal expression to the prevailing seventeenth-century accounts of affect, cognition, and perception should perhaps prompt us to reconsider this division of labor, with its implicit assumption that graphic and verbal forms of representation produce incommensurate experiential outcomes. In this book's later chapters, we shall see that many philosophers of the seventeenth century had a starkly different conception of the capacities of verbal and graphic representation. Grew and his contemporaries believed that well-chosen words could sometimes produce the same mental effects—the same vivid and striking mental images—as pictures themselves. They saw verbal descriptions as repositories of images, comparable both in function and in anticipated cognitive effects to the objects that we would recognize as having visual properties. Meanwhile, they often thought about pictorial representations in rhetorical terms, valuing them for their capacity to bring forth the affective states associated with imaginative picturing in poetic and rhetorical theory. Philosophers such as Boyle, Grew, Hooke, Ray, and Willis understood the capacities of verbal and graphic representation in unfamiliar terms, picturing the world through experiential categories that do not neatly align with our own tools for investigating the affective properties of words and images.

Aesthetic Science therefore proposes an approach that seeks to historicize not just words and images but a regime of experience. As Ofer Gal and Raz Chen-Morris have made clear in their book *Baroque Science*, this end can only be accomplished by synthesizing the remarkably wide range of disciplines and practices at stake when philosophers of the seventeenth century set out to reckon with the world through their senses.[45] In the early Royal Society, the techniques of graphic and verbal representation were central to the empirical project, and the analytical tools of visual and literary studies provide us with powerful tools for understanding them. We can only reckon with the affective work expected from those representations, however, by recognizing contemporary concerns about the effects of sensory experience on the mind and body, taking in fields including not just rhetoric and poetics but also the neurosciences, treatises on the passions, theology, and medicine. For many in the Royal Society, moreover, the metaphysical presuppositions and aesthetic claims of the design argument had an important role. Questions about design in nature had implications for virtually all the practices at stake in the empirical project, ranging from apparently prosaic questions about the practices of visual representation to complex debates about the extent to which the material organs of sensation could yield reliable insights into the nature and attributes of an imperceptible, immaterial God. Indeed, one of the key claims of this book will be that some of the Royal Society's leading members drew close parallels between the affective states arising from the encounter with design in nature and the affective states associated with clear and distinct mental picturing. Philosophers such as Boyle, Grew, Hooke, Ray, and Willis could thus use the resources of rhetoric and poetics to produce a form of affective intersubjectivity. Mobilizing strategies of vivid imaging common to texts and pictures alike, they sought to give their readers the capacity to experience forms of pleasure correspondent not only to the possession of clear ideas but also to the recognition of God's involvement with the affairs of the world.

The Approach

This book focuses, as we have seen, on a group of natural philosophers who argued that the close investigation of natural things would reveal their design and consequently also God's close relationship to his creation. Rather than offering a comprehensive account of the aesthetics of empiricism, *Aesthetic Science* explores the way in which philosophers such as Boyle, Grew, Hooke, Ray, and Willis enacted one version of the design argument in the pursuit of knowledge through sensory experience. Although they will not be our focus, we will thus sometimes encounter contemporary philosophers—including others in the Royal Society—who did not agree that the design seemingly to be found in nature was a compelling source of theological knowledge. In addition, we will not deal extensively with the version of the design argument promoted by Newton and his followers from the turn of the eighteenth century onward. As we shall see, this version of the design argument was compatible with the physico-theology practiced by Ray and his contemporaries, especially in its emphasis on the pleasure to be taken in identifying the regularities on display in the motions of the celestial bodies and its insistence on the insufficiency of purely mechanical explanations. Given that it was not based on the close sensory investigation of material things, however, it has far less to teach us about how natural philosophers reckoned with the affective dimensions of experience. Although its framework was broadly teleological, it did not depend to the same extent on defining the precise purposes for which natural things had been designed. The stars and planets were too far away, and thus too inscrutable, for their purposes to be known with the same kind of precision to be found in plants and animals. As well as being very well known, the Newtonian version of natural philosophy therefore gives us fewer insights into how natural philosophers used empirical strategies to reason about design in nature.[46]

While recognizing such differences, the aim of this book is to focus on a single, relatively coherent, group of natural philosophers. The advantage of this approach is that it creates a space for the interdisciplinary breadth needed to reconstruct the affective dimensions of that group's brand of empiricism. In an argument consisting of three parts, we shall see that Boyle and some of his key contemporaries shared a commitment to experiencing the world as the product of design. Tracing that commitment through their metaphysics and theology, models of cognition, and strategies of inquiry and representation, we can bring the wide range of ideas and practices at stake in their empirical project into a single focus.

The first part of the argument, taking in chapters 1 and 2, reconsiders the design argument from the perspective of its relationships to both the explanatory frameworks and the representational practices of natural philosophy. For a long time, scholars have regarded expressions of the design argument such as Ray's *Wisdom of God Manifested in the Works of Creation* (1691) as apologetic works, reflecting a desire to deflect accusations of religious impropriety rather than the metaphysics or aesthetics of empiricism. Chapter 1, "Physico-Theology, Natural Philosophy, and Sensory Experience," reconsiders this view through a new interpretation of Ray's *Wisdom of God* and crucial works of natural theology including Boyle's *Disquisition about the Final Causes of Natural Things* (1688). Examining the sources and strategies they used to reveal God's involvement with the mechanics of nature, this chapter will show that Ray and Boyle saw the design argument as a form of empirical theology, mobilizing explanatory strategies taken from their natural philosophy. Far from simply recycling a well-worn commonplace of natural theology, Ray and Boyle framed a design argument that depended on highly specific claims about the kinds of causation that could be made accessible to sensory experience and the strategies appropriate to representing them. Chapter 2, "An Empiricism of Imperceptible Entities," examines even deeper links between the design argument and natural philosophy by focusing on their shared preoccupation with making imperceptible entities—both infinitesimally small atoms and an infinite, immaterial God—accessible to sensory experience. Culminating in a discussion of Boyle's attempt to make corpuscular particles and divine wisdom simultaneously perceptible in a chemical demonstration of the possibility of the resurrection, this chapter reveals that he and his contemporaries encoded the ontology of a world designed by God into both their representational practices and their accounts of the workings of the brain and senses. The first part of the book seeks to show, therefore, that design structured the ways in which philosophers reckoned to perceive, represent, and explain natural things.

In its second part, *Aesthetic Science* moves from the theology and metaphysics of design to the practicalities of finding and representing the agency of a designer in both natural and artificial things. Taking up the length of chapter 3, this part of the argument challenges the view that Hooke and others in the Royal Society practiced a form of empiricism aiming at the accurate transcription of experience into either graphic or verbal form. Entitled "In Search of Lost Designs," this chapter explores the challenges that arose when philosophers believed themselves to be working with objects that did not exhibit the signs of beauty and perfection to be expected from the work of a talented designer, whether human or divine. We shall ask, in other words, what happened when philosophers believed themselves to be confronted by the ruins of natural and artificial bodies that had once exhibited a far greater degree of perfection. Starting with a new interpretation of the seventeenth-century debate about the original appearance of Stonehenge, I explore the strategies used by naturalists to recover the intentions of a designer from ruined forms that did not easily reveal their purposes. Comparing such architectural debates to the practices at stake in the interpretation of apparently ruined bodies such as snowflakes, stuffed birds, and the roots of plants, we shall see that the aesthetics of design played a surprisingly important role in the everyday practices of natural philosophy. Drawing on sources such as Hooke's verbal and graphic depictions of crystals produced by freezing, and a grumpy exchange of letters between Grew and the naturalist Martin Lister (ca. 1639–1712) about the anatomy of vine roots, this chapter reveals the extraordinary lengths to which philosophers sometimes went to find pleasing examples of design in nature—even when that design was by no means obvious to the senses.

The final two chapters seek to understand the pleasures associated with the experience of design. Chapter 4, entitled "Verbal Picturing," offers a new interpretation of the verbal descriptions used by naturalists as the primary medium for representing natural things. Focusing on the work of Ray, perhaps the most indefatigable describer at work in the early Royal Society, this chapter reconsiders the relationship between words and pictures in the formation of mental images. Drawing both on the descriptions themselves and on overlooked theoretical works such as Ray's dissertation on different approaches to plant classification, *De Variis Plantarum Methodis Dissertatio Brevis* (1696), we shall see how Ray and his contemporaries believed that verbal descriptions could lead to the formation of pleasing mental images. Indeed, they valued pictures—whether produced graphically or verbally—chiefly for their capacity to elicit the cognitively beneficial pleasure associated with the possession of clear and distinct ideas. Taking up this insight, chapter 5, "Natural Philosophy and the Cultivation of Taste," asks exactly what Ray and his contemporaries hoped to accomplish with their descriptions. Exploring the connections between rhetoric and neurology in works such as Boyle's *Some Considerations Touching the Style of the H. Scriptures* (1661) and Willis's *Cerebri Anatome*, this chapter shows that verbal description served purposes that went far beyond representation. Confronted by the recognition that not everybody took as much pleasure in the study of nature as they hoped, Ray and his contemporaries positioned verbal description as a means of resolving what we could call a problem of taste. Exploiting the power of figurative strategies such as the comparison, they thought it possible to gradually give their readers the capacity to take pleasure in the encounter with divine design. They saw description as a tool not only for conveying information but also for transmitting the affective or subjective states on which the production of knowledge through sensory experience depended.

Insisting that the history of empiricism can only be told through a history of experience, *Aesthetic Science* reveals a decidedly premodern kind of knowledge production. Far from seeking simply to picture the world in the most accurate way possible, Ray and his contemporaries closely identified the discovery of the truth with the cultivation of the capacity for affective states that, from our perspective, look rather subjective. They sought to generate a community of feeling subjects, united by a kind of intersubjective agreement about what it felt like to experience a world designed by God. Such an insight may permit us to reconsider not only the history of the empirical sciences but also the genealogy of aesthetics as it would emerge over the course of the eighteenth century. The notion that the empirical sciences depended on anesthetizing the mind and body against the affective dimensions of experience fits all too easily with an account of aesthetic experience that disconnects the pleasures of beauty from the work of reasoning about the world. By recognizing the affective dimensions of empiricism in the seventeenth century, we may thus begin to pose broader questions about the history of obtaining knowledge through our sensory encounters with the world.

1

Physico-Theology, Natural Philosophy, and Sensory Experience

From its earliest origins, the Royal Society of London promoted an empirical approach to natural philosophy. The study of natural phenomena accessible to sensory experience was to provide the foundation for a reformed science of nature. Some of the Royal Society's leading members, including Robert Boyle, Nehemiah Grew, and John Ray, argued that a similar approach could also serve theological purposes. They argued that the study of nature could lead them beyond knowledge of physical phenomena toward a deeper understanding of the God who had supposedly brought them into being. What they sought, in other words, was a theology based on the same evidence as their natural philosophy. The result was to be a distinctive form of natural theology based on the discovery of design in natural things—one that came to be known as "physico-theology."

Understood in broad terms, natural theology in the Western tradition is extremely ancient. It has a played a role in Christian theology since the days of the earliest church fathers, and its roots are to be found in the thought of classical antiquity. Its aim has always been to work out what can be learned about God through human reason alone—without the supernatural aid of scriptural revelation or spiritual inspiration.[1] In the seventeenth century, therefore, theological authors almost invariably identified a sharp distinction between "natural" and "revealed" religion. Natural religion could, through the exercise of human reason, demonstrate God's existence and even some of God's attributes. The authority of revelation was still needed, however, to persuade the mind to accept those mysterious doctrines—such as the death and resurrection of Jesus—that defied rational explanation. John Wilkins thus defined natural religion as that "which men might know, and should be obliged unto, by the meer principles of *Reason*, improved by Consideration and Experience, without the help of *Revelation*."[2]

Arguments based on the contemplation of nature have long been important to natural theology. The proofs offered by Stoic philosophers for the existence and activity of the gods, given their most famous airing by Cicero (106–43 BCE) in *De Natura Deorum* (*On the Nature of the Gods*; 45 BCE), depended to a large extent on examples purportedly demonstrating that order and purposiveness were to be found in the natural world. In Christian natural theology it was more common, however, to mobilize the evidence of nature only alongside—and often in subordination to—a range of other argumentative strategies. This was the case with the influential form of natural theology promoted by philosopher-theologians such as Henry More (1614–1687) and Ralph Cudworth (1617–1688) around the middle decades of the seventeenth century. In common with many of their contemporaries, More and Cudworth put just as much emphasis on the argument that the human soul could deduce the existence of God by contemplating the conditions of its own existence—albeit finding the mode of ontological speculation pursued by René Descartes unsatisfactory—as they did on inferring the same thing from the divine workmanship apparently to be found in nature.[3] The proponents of physico-theology, by contrast, insisted that the study of nature could yield insights far more effectively than any of the other available strategies. In the preface to his *Wisdom of God Manifested in the Works of Creation* (1691), Ray asserted that "Proofs taken from Effects and Operations, exposed to every Mans view, not to be denied or questioned by any, are most effectual to convince all that deny or doubt of [God's existence]." Here, Ray signaled that he would squeeze ontological speculation to the margins of natural theology, focusing instead almost exclusively on proofs based on natural phenomena accessible to sensory experience.[4]

Physico-theology therefore depended on the notion that sensory experience could supply reliable evidence for God's existence and attributes. For Ray and his contemporaries, this evidence was—with a few notable exceptions—to be found in what natural philosophy could reveal about nature's design.[5] Consider Walter Charleton's *The Darknes of Atheism Dispelled by the Light of Nature* (1652), a work that has come to be regarded as the first in the English physico-theological tradition. There, Charleton asserted that the exquisite design to be found in even the minutest of animal bodies could evince God's role in the world:

> Since [. . .] the *proboscis* or trunk of a *Flea* [hath] more industry in its delicate and sinuous perforation, then all the costly *Aquæducts* of *Nero's Rome* [. . .]

> how can it be that man, noble and ingenious man, should [. . .] admit the managery of an *Architect*, or knowing principle, in the structure of a *house*, and yet determine the more magnificent Creation of the *Universe* upon the blind disposal of *Fortune*?[6]

Like those following him later in the century, Charleton made two overlapping claims here about how sensory experience could lead to knowledge of God. The first was that close inquiry would reveal natural things to be the products of divine design, manifesting the same kind of purposiveness to be found—note his architectural comparisons—in the products of human design. Second was the notion that encounter with this design would give rise to some form of pleasure, expressed in this case by Charleton's unsettling admiration for the "delicate and sinuous" form of the flea's proboscis. For the physico-theologians, close investigation would reveal nature to be the material manifestation of God's wisdom and purposes, both instructive and delightful to behold.

The physico-theologians made this argument, however, in the context of widespread concern that the pursuit of natural philosophy might in fact be harmful to religion. Boyle and Ray identified the materialist philosophies of Thomas Hobbes and Baruch Spinoza as the most serious threat, regarding such efforts to explain causation in purely physical terms as prejudicial to the notion that God was actively involved with the natural world. They also took issue with René Descartes, admonishing him for acknowledging an immaterial God as the world's maker without at the same time showing that such a God was also required to sustain that world after his initial creative act. Seeking to respond to these threats, Ray and his contemporaries positioned physico-theology as a form of religious apologetic, emphasizing what they took to be the simplicity and affective force of its appeals to nature's design and the pleasure to be had in encountering it. Such easy and obvious arguments, they asserted, were ideal tools both for refuting atheism and for demonstrating that their own approach to the study of nature was ideally placed to reinforce conventional religious doctrines. Since the 1980s, therefore, historians have for the most interpreted physico-theology as a strategy for justifying the pursuit of natural philosophy to a concerned public. Neal C. Gillespie has argued, for instance, that Boyle and Ray saw physico-theology as a powerful tool for promoting unity among different groups of Protestants in the wake of the religious conflicts bound up with the English Civil War. More recently, Scott Mandelbrote has portrayed it as a response to criticisms of the state of university education that arose during the 1650s.[7]

The problem with these arguments is not that they are incorrect but rather that they offer too narrow a perspective, regarding physico-theology solely through the lens of its polemical functions. Indeed, they give the misleading impression that Ray and his contemporaries saw the design argument as little more than a tool of polemical debate, useful for reinforcing conventional religious belief but subject to much looser standards of proof than natural philosophy. They suggest, therefore, that physico-theology had little to do with the forms of knowledge production at stake in natural philosophy, serving instead as a simple and straightforward means of reinforcing conventional religious beliefs.[8] In this sense, recent interpretations echo the outlines of an argument made, albeit for different reasons, by an older generation of scholars. In his *Science and Religion in Seventeenth-Century England* (1958), for example, Richard S. Westfall sought to explain how—as he saw it—Ray and his contemporaries could use a recognizably modern form of empiricism to study nature but nonetheless squeeze what they found into the antiempirical framework of the design argument. His response was to argue that scientists of the late seventeenth century were caught between two contradictory mentalities: a genuinely empirical impulse to record the truth about things and a religious belief powerful enough to make them overlook facts that might otherwise have prevented them from seeing the world as a harmonious product of divine wisdom. While he thus had far less to say about physico-theology's polemical functions, he nonetheless mobilized a similar distinction to explain its relationship with natural philosophy. Westfall saw the design argument as a matter not of knowledge but rather of belief, without much of a connection to natural philosophy and its strategies for understanding the physical causes of natural phenomena.[9]

I do not think, however, that Boyle and his contemporaries could have embraced so strict a demarcation. Peter Harrison and Jonathan Sheehan have both recently produced forceful reminders that seeing theology as a matter of irrationally acquired belief—rather than rationally acquired knowledge—is a modern idea. It is no great generalization to point out that, during the medieval and much of the early modern period, theology was often regarded as a science, as much capable of producing demonstrative truths as philosophy. During the seventeenth century, however, the scientific status of theology came under pressure, with thinkers such as Descartes, Hobbes, and Spinoza seeking to exclude an increasingly large number of theological questions from the domain of rational inquiry.[10] Such efforts to make theology into a matter of mere belief met, however, with fierce resistance—not least on the part of many affiliated with the Royal Society. To pigeonhole the physico-theology of Boyle, Grew, Ray and others as a matter of belief alone would thus be to take sides in what was at the time an unresolved debate. It is also worth recalling that thinkers of the seventeenth century did not generally share today's belief that the sciences are incapable of providing meaningful answers to metaphysical questions. As Quentin Meillassoux reminds us in his recent book *After Finitude: An Essay on the Necessity of Contingency*, metaphysics became a matter for religious belief only after the interventions of David Hume and Immanuel Kant in the second half of the eighteenth century. By contrast, philosophers of the seventeenth century saw questions about the fundamental conditions of existence and the ontological status of both material things and spiritual beings as legitimate targets for philosophical investigation.[11]

The following chapter will take a new approach to the interpretation of physico-theology. Rather than starting with the assumption that physico-theology was ultimately a matter of belief, we shall entertain the suggestion, made repeatedly by Ray and his contemporaries, that it was an exercise in the production of knowledge. For much of this chapter, as a result, we shall compare Ray and Boyle's strategies for evincing the design of nature with those to be found in the older theological texts on which they drew, both ancient and modern. Only through such a comparison can we work out whether they simply made expedient use of theological arguments that already existed to reinforce conventional religious beliefs, or whether they mobilized representational and explanatory strategies derived from their natural philosophy to push those arguments in a new and distinctive direction. The intention here, however, is not simply to settle questions about the connections between physico-theology and natural philosophy, nor even to show that today's distinctions between the domains of knowledge and belief are unhelpful when it comes to thinking about the relationships between science and religion in the seventeenth century. Rather, our purpose is to address an even more fundamental question that can shed light on the interrelationships between science and religion. The question, that is, of what it meant to use the organs of aesthetic perception as instruments in the production of knowledge, whether concerning the physical causes of natural things or the ultimate reasons for their existence. By examining how Ray and his contemporaries set out to make nature's design, and by extension the wisdom of God, accessible to the senses, we can begin to see how they thought it possible to answer metaphysical and theological questions by empirical means. By showing, moreover, that physico-theology was an exercise in the production of knowledge, we can begin to question whether its rhetoric about the moral and affective dimensions of the sensory encounter with design in nature were relevant to the practices of natural philosophy.

Final Causes and Natural Philosophy

Works of both natural philosophy and theology containing physico-theological arguments were published throughout the second half of the seventeenth century. It was not until the latter part of the century, however, that physico-theology emerged as a widely recognized genre of theological persuasion, with claims about design deployed to the exclusion of other forms of reasoning about natural phenomena. Perhaps the most important single contribution to the genre, as I indicated in the introduction, was Ray's *Wisdom of God Manifested in the Works of Creation*, published in 1691. Ray framed the work as an extended sermon, using a straightforward style and encouraging tone to make physico-theology accessible to a relatively wide audience. The book was to prove both influential and popular, providing a pattern for physico-theologians to follow in the future. From 1691 to 1743, *The Wisdom of God* was published in eleven editions by London booksellers alone, and it also appeared in French, German, and Dutch versions.[12] Although Boyle wrote, as we shall see, the only contemporary theoretical account of how to reason about design in nature, his role in shaping the new genre of natural theology was less direct. In 1691, the last year of his life, he added a codicil to his will leaving funds for an annual series of lectures dedicated to "proving the Christian Religion, against notorious Infidels, viz. Atheists, Theists, Pagans, Jews, and Mahometans, not descending lower to any Controversies, that are among Christians themselves."[13] In the years following Boyle's death, clergyman-philosophers including Richard Bentley and Samuel Clarke honored this bequest by delivering and publishing sermons that promoted the study of nature as a means of learning about God's existence and attributes. Among them was William Derham (1657–1735), who served as the principal custodian of Ray's intellectual legacy after the latter's death in 1705. He modeled his own *Physico-Theology; or, A Demonstration of the Being and Attributes of God, from his Works of Creation* (1713) on Ray's *Wisdom of God*.[14]

In keeping with their desire to speak to a broad audience, the physico-theologians often sought to give the impression that learning about God through the discovery of design in nature would be easy. Ray opened *The Wisdom of God*, for instance, by declaring that the signs of God's wisdom in nature were so obvious that "illiterate Persons of the lowest Rank of the Commonolty [*sic*]" could recognize them. Derham would repeat this point in his *Physico-Theology*, announcing that nature's contrivances were "the most easy and intelligible Demonstrations of the *Being* and *Attributes* of God," easily understood by "such as are unacquainted with the Subtilties of Reasoning and

Argumentation."[15] These claims, repeated throughout the physico-theological corpus, had the obvious goal of encouraging readers to accept as easy and self-evident the argument that studying nature could yield worthwhile insights into God's wisdom and intentions. At the same time, they had the convenient effect of obscuring the fact that the key claims of physico-theology were the subject of ongoing debate and disagreement. Both within and without the Royal Society, there were natural philosophers who did not find the suggestion that the discovery of apparent design or purposiveness in nature could yield reliable knowledge of God's existence and attributes very convincing. To understand exactly how Ray and his contemporaries thought it possible to obtain theological knowledge from the empirical study of nature, we must therefore avoid being taken in by their attempts to portray the enterprise as simple and obvious. Instead, we should begin by briefly surveying the seventeenth century's debates about the place of teleological reasoning—reasoning about things in terms of their supposed purposes—in natural philosophy.

During the seventeenth century, philosophers such as Galileo, Descartes, and Boyle proposed mechanical models of causation, seeking to explain natural phenomena as the result of collisions between otherwise inert particles of matter. The only physical causes they claimed to admit were those arising from the material agency of matter in motion. Such mechanical models contrasted sharply with the forms of explanation used by practitioners of the broadly Aristotelian natural philosophy then still dominant in the universities and known today as *Scholasticism*. The Scholastics generally admitted what Aristotle had called "final causes"—the purported purposes for which things existed—into their explanations. A Scholastic thinker could explain the fact that heavy bodies generally fell to the ground by supposing that downward motion fulfilled one of the purposes for which those bodies existed. As we shall see further on, Descartes and Boyle did not agree about whether teleological reasoning of this broad sort was incompatible with a mechanical account of physical causation. They were unanimous, however, in their view that the Scholastic approach to understanding how material things acquired and fulfilled their purposes was a major problem. In his *Physics*, Aristotle had sought to explain how natural things change over time by suggesting that they had not only a material component but also an intellectual one that he called the "form." The forms, he argued, were entities that organized matter and made it capable of acting toward rationally intelligible purposes. Take, for instance, the fact that sunflower seeds have an observable tendency to develop into sunflowers. Following Aristotle's reasoning, this regular and seemingly purposive process could be explained by the form's capacity to make the matter to which it was attached move toward the fulfilment of a specific idea—that of becoming a sunflower. Scholastic thinkers thus held that purposiveness in nature did not result from straightforwardly material causes. It could sometimes result, they supposed, from the immanent agency of so-called substantial forms that somehow made matter work toward given ends.[16]

Proponents of the mechanical philosophy took two distinct approaches when attacking this form of teleological reasoning. The first was to argue that—as immaterial agents that could be neither perceived nor even imagined—substantial forms were completely unintelligible and consequently incapable of shedding much light on how or why things happen. Both Descartes and Boyle, for example, unfairly insisted that even the wisest among those who admitted substantial forms as explanations could say little about what they really were or how they imparted purposes to material things.[17] The second line of attack was to cast Scholastic final causes as hopelessly anthropocentric, attributing human purposes to things that had not necessarily been made with such purposes in mind. Francis Bacon (1561–1626), whose efforts to reform the methods of natural philosophy were a crucial source of inspiration for many in the Royal Society, pronounced in his *New Organon* (1620) that "final causes, which are plainly derived from the nature of man rather than of the universe [. . .] have wonderfully corrupted philosophy."[18] Writing later in the century, the physico-theologians sought to dismiss the scholastic account of final causes by equating it with the far less subtle and far more blatantly anthropocentric argument that the world simply existed for the sake of its human inhabitants. Both Ray and Boyle, for instance, painted the notion that the stars existed only to shine light on the earth—or to provide navigational assistance—as an obvious absurdity. As John Hedley Brooke has wisely pointed out, however, their rejection of the idea that the world was made only to meet human needs was not an outright rejection of the notion that nature served purposes that humans could work out. Despite rejecting the Scholastic position, indeed, Boyle was perfectly content to entertain the possibility that human ends were a small subset of the grander purposes for which God had set the heavens in motion.[19]

By the middle of the seventeenth century, therefore, it was by no means obvious that the resources of natural philosophy could be used to learn much about the final causes of natural things. Unlike Bacon, however, Ray and his contemporaries nevertheless aimed to show that the study of nature could yield useful and reliable knowledge of God's intentions. In doing so, they sought not only to respond to those who reasoned incorrectly about final causes but also to answer the acute challenge posed by those philosophers who denied that nature had any final causes at all. Troublingly, this challenge appeared to come principally from other proponents of the mechanical philosophy. Thomas Hobbes (1588–1679) had argued, for instance, that an account of causation based on purely mechanical interactions between imperceptible particles of matter could on its own, without recourse to any immaterial entity, explain the appearance of things. For Ray and most of his contemporaries, suggesting that the world could get along without the guidance of an intelligent, immaterial agent was tantamount to atheism. Indeed, they tended—simplifying a much more complex intellectual lineage—to equate materialism of the Hobbesian sort with the famously atheistic hypothesis of the ancient philosopher Epicurus (341–271 BCE), best known to posterity through Lucretius's first-century BCE poem *De Rerum Natura* (*On the Nature of Things*).[20] Epicurus had argued that the world was too chaotic to be the product of wisdom, offering instead an account of physical causation based on happenstance collisions between randomly moving atoms. Arguing that the world and everything in it had arisen from chance interactions between imperceptible particles of matter, he could argue that the gods were not at all involved with its affairs. The Epicurean hypothesis thus exemplified the danger posed by any attempt to give a purely material account of physical causation. To argue that the workings of nature could be explained without recourse to final causes imposed by an immaterial, intelligent agent was to call God's existence and active engagement with the world into question.[21]

Ray and his contemporaries therefore worried not only about the outright materialism of thinkers like Hobbes but also about the dualist mechanical philosophy associated with Descartes. Unlike Hobbes, Descartes had argued that the world required an infinite, immaterial God for its initial creation. Descartes had gone on to suggest, however, that the world had no need of divine intervention thereafter, picturing the world—including all animals and even the functions of the human body—as a gigantic clockwork machine quite capable of unfolding a complex series of operations without any further guidance. As we shall moreover see shortly, Descartes also denied that knowledge of this machine obtained through natural philosophy could yield any useful insights about God's purposes for bringing it into being. The problem here, as Boyle and Ray both saw it, was that such an argument would deny naturalists recourse to the most obvious and effective response to the perceived threat of philosophical materialism. In *The Wisdom of God*, borrowing his words from Ralph Cudworth's *The True Intellectual System of the Universe* (1678), Ray expressed bafflement that Descartes could hypothesize a world governed by the orderly and intelligible laws but still refuse to admit the possibility of learning about God's reasons for creating it:

> But our *mechanick Theists* [followers of Descartes] will have their Atoms [. . .] to have taken up their places and ranged themselves so orderly, methodically and directly; as that they could not possibly have done it better, had they been directed by the most perfect Wisdom. Wherefore these *Atomick Theists* utterly evacuate that grand Argument for a God taken from the *Phaenomenon* of the Artificial frame of things.[22]

Confronted by the materialist assertion that the world had no need of an intelligent creator, the physico-theologians could not treat final causes as a matter of mere belief or opinion. Nor could they be satisfied with a demonstration that the world functioned according to orderly and intelligible principles. As they saw it, the only adequate response to the claims of Epicureanism was an intellectually credible demonstration that the world depended on the intentions of a wise and benevolent God for both its creation and its sustenance.[23]

John Ray and the Representation of Design

In constructing a response to the threat of Epicureanism, the physico-theologians appropriated and adapted resources from an equally ancient tradition of arguments—this time in favor of divine involvement with the world. With examples in both Platonic and Stoic thought, arguments in this tradition purported to demonstrate the activity of the gods in the world by means of analogy between the productions of nature and those of human art. In the second book of *De Natura Deorum*, for instance, Cicero repeatedly asserted that the wisdom and purposiveness on display in the products of human art was obvious to anyone who considered them. Pointing out the many apparently similar examples of such purposiveness on display in nature, whether in the regular motions of the stars or the admirable adaptation of the human ear to its auditory function, he invited readers to suppose that a similarly wise agent must also have brought them into being.[24] John Ray and his contemporaries found Cicero's account of the Stoic teleological argument helpful in several key respects, drawing on it extensively as they framed the new physico-theological genre. For one thing, the use of comparisons between art and nature made it possible to talk about purposiveness in nature without resorting to the Scholastic account of how that purposiveness was accomplished. To imagine God as a supremely accomplished artisan was to imagine a world in which purposiveness was imposed on nature through its design rather than emanating from a mysterious innate intelligence. At the same time, Cicero's design argument could serve as a model for the attempt to appeal to empirical methods. The second book of *De Natura Deorum* seemed to suggest that evidence accessible to the human senses—whether in natural phenomena or the things to which they were compared—was enough material for an intelligible and persuasive account of final causes in nature.[25]

It is thus clear that the physico-theologians turned to the resources of the humanist intellectual culture in which they had all been educated when piecing together their new genre of theological persuasion. I cannot agree, however, with Dmitri Levitin's recent suggestion that physico-theology was thus fundamentally a product of what he calls "traditional humanist concerns," only superficially grounded in new approaches to the production of knowledge about nature.[26] Although this argument arises from a different intellectual perspective, it nevertheless reproduces a much older—and now discredited—line of thought that regards physico-theology as nothing more than a continuation of the Stoic tradition. This line of thought can be traced to the second half of the eighteenth century, when philosophers such as David Hume and Immanuel Kant sought to discredit all such arguments for the existence of God. Working decades after Ray and his contemporaries had died, they used terms such as "natural religion" and "physico-theology" to account for virtually any proof based on the analogy between purposiveness in art and that supposedly on display in nature. In the *Critique of the Power of Judgment*, for instance, Kant described physico-theology as an attempt "to infer from the *ends* of nature (which can be cognized only empirically) to the supreme cause of nature and its properties," supplying a definition that might just as well apply to the scholastic account of teleology as the one that Boyle wanted to put in its place. For better or worse, as Peter Harrison has pointed out, modern historians and philosophers of religion have often adopted the same generalizing perspective, treating the design argument promoted by Robert Boyle and Ray as just one more expression of an older teleological proof that had changed little since antiquity.[27] This interpretation tends to reinforce the view that physico-theology was primarily an exercise in justification, mobilized to persuade the public that natural philosophy was no threat to conventional religion but little connected to its practices for producing knowledge.

To see how Ray could use ancient authorities in the service of a new and distinctive version of the design argument, closely allied to natural-philosophical practices, we must take a little time to consider how he composed *The Wisdom of God*. The book began life as a series of sermons that Ray delivered at the end of the 1650s, while still a fellow of Trinity College, Cambridge. Revising it for publication in 1691, he added many of his own observations on natural history, along with insights gleaned from books released in the intervening years by authors such as Boyle, Ralph Cudworth, and the physiologist Richard Lower (1631–1691). Perhaps Ray's single most important modern source, however, was an influential work of natural theology to which he must have had access when writing the original sermons—*An Antidote against Atheisme*, written by the Cambridge philosopher Henry More and published in 1653.[28] Ray had little use for the *Antidote*'s first and third books, which set out respectively to prove God's existence with an ontological argument, and through a collection of histories testifying to providential interventions in human affairs by spiritual agents. In the second book, however, More sought evidence for God in the study of nature, devoting several chapters to the apparent beauty, order, and purposiveness of natural bodies. In those chapters, More deployed a range of ancient and modern sources, citing authorities such as the Italian mathematician Girolamo Cardano (1501–1576) and the Venetian scholar-physician Julius Caesar Scaliger (1484–1558), while at the same time making extensive but unacknowledged use of Cicero's *De Natura Deorum*.[29] For his own part, Ray in turn made liberal use of More's arguments, examples, and even organizational conceits in *The Wisdom of God*.

Consider how Ray used More's account of the human eye. In the *Antidote*, More opened his discussion with a loose translation of a list of safeguards for the eye's protection taken from *De Natura Deorum*. More adapted Cicero's remarks on the eyelids in the following manner:

> The *Eye-lids* are fortify'd with little stiffe *bristles* as with *Palisadoes*, against the assault of Flyes and Gnats, and such like bold *Animalcula*. Besides the *upper-lid* presently claps down and is as good a fence, as a *Portcullis* against the importunity of the Enemy:[30]

More's paraphrase captures Cicero's meaning reasonably well, but it includes much that is not to be found in the original and that could never be justified as a translation. Cicero had said nothing, for instance, about the role of the eyelids in repelling insects and tiny animals, instead referring only—and in a mere three words—to the threat posed by "any falling object" ("si quid incideret").[31] More did not stop, however, at modifying *De Natura Deorum* to make its imagery more vivid and affecting. As soon as he had done with Cicero's enumeration of its defenses, he added a slightly longer list—taking up about a page—of six observations on the eye's internal workings. Based on recent optical and anatomical research, those observations included the way the uvea's dark color prevents light from bouncing indiscriminately around the inside of the eyeball and the way the ciliary processes enable the eye to adjust its length to focus on objects at different distances. More thus backed up Cicero's remarks on the eye's exterior parts with new evidence purporting to show that its interior must also be a work of intelligent design.[32]

Ray based his discussion of the eye in *The Wisdom of God* on the one in More's *Antidote*. Here, his use of Cicero's remarks on the eye's defenses is telling. Despite including the original Latin text in italics, Ray lifted his ostensible translation from More's paraphrase, retaining not only More's additions to the account of the eyelids but also verbal shibboleths such as the eccentric substitution of "palisadoes" for "palisades."[33] More important, he also modeled his account of the eye's interior anatomy on the one given in the *Antidote*, even preserving—in the same order—the same sequence of six observations. Here, however, the similarities end. Although Ray sought to account for the same relationships between form and function in the structures of the eye, he went about describing them in a strikingly different manner. Consider the contrast between his description of the action of the ciliary muscle and the one left by More. In the *Antidote*, More briefly explained that "the *Tunica Arachnoides*, which invellops the *Crystalline Humour* by vertue of its *Processus Ciliares* can thrust forward or draw back that precious usefull part of the Eye, as the neernesse or distance of the Object shall require."[34] In *The Wisdom of God*, meanwhile, Ray described the muscle's workings at much greater length:

> Because the Rays from a nearer and from a more remote Object do not meet just in the same distance behind the Crystalline Humour [the lens] (as may easily be observed in lenticular Glasses [artificial lenses], where the point of concourse of the Rays from a nearer Object is at a greater distance behind the Glass, and from a further at a lesser) therefore the *ciliary processes*, or rather the ligaments observed in the inside of the Sclerotick Tunicle of the Eye [. . .] do serve instead of a Muscle, by their contraction to alter the figure of the Eye, and make it broader, and consequently draw the *Retine* nearer to the Crystalline Humour, and by their relaxation suffer it to return to its natural distance according to the exigency of the Object, in respect of distance or propinquity.[35]

Before turning to the substance of this description, it is worth noting that Ray subjected the remaining observations to the same treatment. As a result, the part of *The Wisdom of God* based on the *Antidote*'s six observations on the eye is easily three times longer, running to about a thousand words rather than just three hundred.[36]

More's brief description was little more than a statement of the ciliary muscle's function, appearing in a list of similarly brief statements about other parts of the eye. Like Cicero before him, More for the most part listed examples of form and function, heaping one on another to produce a dazzling impression of the immense quantity and variety of God's ingenuity. In much of *The Wisdom of God* Ray used the same rhetorical ploy, and for much the same purpose. When he came to important examples such as the eye, however, he slowed down to mobilize quite another style of description. In his account of the ciliary muscle, Ray began not by pointing out the muscle's function but rather by setting out the optical problem at stake, explaining how the beams of light coming from an object placed close to a lens will meet at a point farther behind that lens than those coming from an object placed farther away. Only then did he move on to describe how the form and action of the ciliary muscle solved that problem as it affected the eye, changing the shape of the eyeball (and thus the position of the retina in relation to the lens) to enable it focus on things both near and far away.[37] The effect of this description was to do something more than merely point out that the ciliary muscle existed and served a useful purpose. Invoking a comparison between lenses of human manufacture and the lens in the eye, it presented an account of exactly how the ciliary muscle served its purpose. While drawing on More's *Antidote*, therefore, Ray took a markedly different approach to the representation of design, seeking not only to point out purposiveness in nature but also to explain exactly how it worked.

This tendency to turn brief accounts of form and function into detailed depictions of the mechanics of design became more and more pronounced in the years to come, both in Ray's works and those published by his followers.[38] As soon as *The Wisdom of God* came out, Ray was hard at work producing detailed supplements to the work. In the second edition of 1692, to take just one example, he enlarged what had been a brief remark on the protective purpose of the nictitating membrane (the horny, translucent inner eyelid found in creatures such as birds, cats, dogs, and fish) by adding a detailed account of its mechanism culled from the recently published English translation of the Académie Royale des Sciences' *Mémoires pour servir à l'histoire naturelle des animaux* (1671–76).[39] Concerned that some readers might lose patience with this rather long description—it ran to about eight hundred words—Ray quoted the justification given in the original text for presenting it nonetheless:

> The Particularities of the admirable Structure of this Eye-lid [the nictitating membrane], are such things as do distinctly discover the Wisdom of Nature, among a thousand others, of which we perceive not the Contrivance, because we understand them only by the Effects, of which we know not the Causes; but we here treat of a Machine, all the parts whereof are visible, and which need only to be lookt upon, to discover the reason of its Motion and Action.[40]

In these borrowed words, Ray registered dissatisfaction with an account of final causes limited only to the observation of their effects. He proposed instead that observers could in some cases understand precisely how final causes came about, premising that understanding on two conditions. First, he indicated that the workings of final causation could be made intelligible by being made accessible to sensory experience, whether through direct witnessing or vicariously through the provision of a detailed description. Second, he indicated that those contrivances could be thought to operate in the same way as those imposed on the products of art by human designers.

Taken in a broad sense, the design argument had always depended on an analogy between nature and art. For the most part, however, More and Cicero had pursued that analogy at a very general level, seeking only to point out that nature exhibited roughly the same sort of intentionality identifiable in the work of a human designer.[41] In *The Wisdom of God*, by contrast, Ray used comparisons between nature and art not only to argue that nature was the product of design but also to explain precisely how that design played out in individual specimens of divine workmanship. With his descriptions, he sought to show exactly how the intricate mechanisms imposed on material things enabled them to fulfill their purposes. Although Ray modeled *The Wisdom of God* on older works of natural theology, he made a design argument that depended on appeals to markedly different experiential and explanatory practices. He purported to make God's methods for accomplishing his purposes accessible to sensory observation. Seeking to make those methods intelligible, moreover, he invoked the guiding metaphor of the mechanical philosophy, inviting readers to imagine natural things as if they depended on mechanisms just like those so readily observable in the products of human art. Unlike More and Cicero, Ray set out to make the mechanisms of final causation into objects of empirical knowledge, mobilizing representational and explanatory strategies derived from his natural philosophy.

An Empirical Theology

In *The Wisdom of God*, John Ray only touched briefly on what was perhaps the strongest objection to any project that aimed to bring final causes back into the domain of natural philosophy. Around fifty years earlier, in the *Meditations* and accompanying *Objections and Replies* (1641), René Descartes had contended that the gulf between the finite human faculties and the infinite mind of God was so great that any attempt to work out God's reasons for making natural things one way rather than another would always run into uncertainty. Knowledge of final causes was impossible, he had suggested, because there was no way to be sure that the purposes assigned by finite human minds to natural things would correspond to the infinite array of purposes that God might have had in mind. As the natural philosopher Pierre Gassendi (1592–1655) recognized in his "objection" to the *Meditations*, such an argument threatened to invalidate any argument for God's existence based on the observation of nature, turning even the simplest observations of functionality in natural things into unverifiable matters of opinion.[42] The same considerations weighed heavily, moreover, with others who were closer to Ray and Boyle. Rhodri Lewis has shown that William Petty, one of the Royal Society's founding members, aired similar doubts in a manuscript treatise written during the late 1670s and given limited circulation thereafter. In that text, Petty suggested that the study of nature could reveal little about God except for the immense distance between his infinite perfections and the imperfections to be found in humans and the fallen world that they inhabited.[43]

It is not clear that Boyle ever knew of Petty's controversial views. In *A Disquisition about the Final Causes of Natural Things* (1688), however, Boyle made a concerted effort to refute Descartes's denial of physico-theology. For Boyle, the crucial problem with Descartes's argument was that it denied sensory experience any role in the generation of knowledge about God's purposes in creating the world. It denied that empirical methods could have much theological use at all. In response, he was compelled to explain precisely how, and under what conditions, the experiential and explanatory practices of empiricism could yield insights about God rather than just the usual hypotheses about physical causation. Boyle thus produced what would turn out to be the only contemporary theoretical work of physico-theology—the only one containing a detailed discussion of how to make the evidence of design in nature into an intelligible representation of the wisdom of God.[44] Despite some important differences between Boyle and contemporaries including Grew and Ray, *Final Causes* must therefore serve as a crucial source for understanding how physico-theology functioned both as a form of knowledge production and as an instrument of theological persuasion. Boyle sought to show precisely how detailed descriptions of the mechanisms of design—just like those offered by Ray in *The Wisdom of God*—could not only demonstrate God's existence but also inspire sentiments of devotion toward God. He argued, in other words, that empirical methods could sustain a uniquely powerful form of natural theology.

Boyle certainly recognized the force of Descartes's argument that it would be presumptuous to think any human could plumb the depths of God's infinite wisdom. In *Final Causes*, he was careful to avoid claiming that natural philosophers could obtain anything like an exhaustive account of all the purposes for which things existed. Following a line of thought begun by Gassendi, he proposed that only a few of God's purposes, identifiable in a limited range of phenomena, lay open to discovery.[45] Characteristically, Boyle made his point using a metaphor, imagining the natural philosopher investigating purposiveness in nature as a half-educated rustic confronted by an extremely complicated sundial:

> Suppose that a Country Man, being in a clear day brought into the Garden of some famous Mathematician, should see there, one of those curious Gnomonick Instruments, that show at once, the *place* of the *Sun* in the *Zodiack*, his *Declination* from the *Æquator*, the *Day of the Month*, the *Length of the Day*, &c. It would indeed be presumption in him, being unacquainted both with the Mathematical Disciplines, and the several Intentions of the Artist, to [. . .] think himself able, to discover all the Ends, for which so Curious and Elaborate a Piece was framed. But when he sees it furnished [. . .] with all the Requisites of a Sun *Dial*, and manifestly perceives the Shadow to mark [. . .] the Hour of the Day; 'twould be no more a Presumption than an Error in him to conclude, that (whatever other Uses the Instrument [. . .] was design'd for) it is a Sun-*Dial*, that was meant to shew the Hour of the Day.

This comparison was not, from a philosophical perspective, an adequate response to Descartes's argument. Descartes had premised the unsearchability of God's wisdom not on its great complexity but rather on its infinity. Nevertheless, it served to make plain Boyle's point that our inability to know all of God's intentions did not have to prevent us from knowing those that lay within the scope of human investigation. Like the sundial obvious even to the uneducated observer of an otherwise baffling gnomonic instrument, some of God's purposes were so "manifest and obvious" that it made no sense to deny them.[46]

Just like Ray, however, Boyle was a little disingenuous in suggesting that the discovery of final causes would be easy. For the most part, he instead put forward explicitly the argument that Ray had made only tacitly through his detailed representations of design. Maintaining that a bare acknowledgment of purposiveness in nature was insufficient, he argued that knowledge of final causes had to include a detailed understanding of the mechanisms by which those causes were brought about. He added, moreover, that such knowledge was only to be obtained through specific experiential and representational practices, in effect restricting the serious study of final causes to natural things that could be subjected to those practices. He therefore went so far as to deny the possibility of learning much about the design of some things that were otherwise of great interest to natural philosophers:

> I cannot but think, that the Situations of the Cœlestial Bodies, do not afford by far so clear and cogent Arguments, of the Wisdom and Design of the Author of the World, as do the bodies of Animals and Plants. And for my part I am apt to think, there is more of admirable Contrivance in a *Mans Muscles*, than in (what we yet know of) the *Celestial Orbs*; and that the Eye of a Fly is, (at least as far as appears to us,) a more curious piece of Workmanship, than the Body of the Sun.[47]

Boyle likely had little doubt that natural philosophers would one day show the stars and planets to be marvelous expressions of divine design. His point, however, was that humans could not presently subject them to close investigation of the sort that would make it possible to obtain detailed knowledge of how they fulfilled their purposes. It was for this reason, then, that he held up plant and animal bodies as far more compelling subjects for the study of final causes. Boyle responded to Descartes's rejection of physico-theology by arguing that only bodies readily accessible to the senses could yield useful insights into the design of nature.[48]

This act of strategic modesty should not, however, blind us to the ambition that it concealed. In both *Final Causes* and the *Christian Virtuoso*, published two years later, Boyle asserted that the experiential and explanatory strategies used by natural philosophers could yield deep insights into the wisdom of God, far surpassing those available to casual observers. We can get a sense of what he meant by considering his advice in *Final Causes* on how to go about uncovering the design of the eye. As Boyle saw it, two relatively specialized operations were required—a dissection that would bring the organ's interior parts into view and an explanation of how those parts worked based on the rules of optics. Together, those two operations would enable an observer not just to see that the eye was intended for the provision of sight but also to grasp the more important point that it was extremely well designed for that purpose, "as the best Artificer in the world could have fram'd a little Engine, purposely and mainly design'd for the use of seeing."[49] In the lengthy discussion that followed later in the work, Boyle thus cataloged the many ways in which human and animal eyes appeared to have been adapted to the laws of optics, noting for instance that the globular lenses found in the eyes of fish were ideally suited to the refractive behavior of light passing through water. He also pointed out that the study of ocular diseases could shed light on the excellence of the eye's design, showing that even the slightest change to the posture or texture of one of its parts would compromise its functionality.[50] Boyle thus argued that his philosophy could deliver far deeper insights into the wisdom displayed in the works of creation than any natural theology that had come before. Only an "Experimental Philosopher," he declared in the *Christian Virtuoso*, had the ability to show precisely "*how wise* an Agent [God] has in those Works express'd himself to be."[51]

Boyle was quite aware, however, that his case would remain unconvincing unless it included a demonstration that the mechanisms identifiable to humans in organs such as the eye could plausibly explain how God transmitted his purposes to nature.[52] Boyle responded to this problem by mobilizing his own attempts, given in earlier works such as *The Origine of Formes and Qualities* (1666) and *A Free Enquiry into the Vulgarly Receiv'd Notion of Nature* (1685), to provide a metaphysical foundation for his laws of nature. That is, he folded his account of how God imposed his intentions on the world into his explanation for how nature could be thought to follow—without itself possessing any intelligence—a divine will that generally expressed itself as a series of orderly, intelligible laws. The first part of this explanation consisted in simply asserting that God had chosen to impose laws governing the interactions between the material constituents of the world—laws that he could choose to upend or suspend whenever he so chose. As Timothy Shanahan has pointed out, this position appealed to Boyle because it enabled him to assert the primacy of divine will over creation (a theological position to which we shall return in chapter 2) and to argue that God's continual intervention was required for the laws of motion to remain in effect.[53] Next, Boyle proposed that God had shaped all the particles of matter in the universe so that, upon imparting motion to them, their collisions and other interactions would lead them to blindly fulfill the purposes that he had in mind for them. Boyle could thus argue that God imposed that purposiveness on matter from the outside, just as a human designer imposed purposes on a clock by giving its lifeless wheels and cogs a variety of shapes and setting them into motion. The design in nature could be thought to arise in the same manner as in the products of art—not from the kind of indwelling intelligence imagined by the Scholastics, that is, but rather from the interactions between particles of matter given various shapes and then made to move with varying quantities of motion.[54]

For Boyle, the main advantage of this kind of explanation was the ease with which it could be understood. As Jan Wojcik has pointed out, Boyle did not generally argue that the mechanical philosophy was intrinsically more plausible than any other but rather that it was the only one to propose an intelligible model for causal agency. For the most part, indeed, he dismissed substantial forms for what he took to be their incomprehensibility, arguing that the human mind could never grasp entities that lay outside the realm of experience.[55] He premised the superiority of his mechanical explanations, meanwhile, on the ease with which the mind could grasp them through sensory experience. In the sensory investigation of artificial machines such as clocks, watches, and automata, he argued, complex purposiveness could be observed to arise from the composition and motion of their material constituents. As Boyle proposed in *Final Causes*, the experience of mechanical causation in the products of art could furnish the mind with a means to explain how natural things accomplished apparently similar forms of purposiveness:

> And when I have seen, as sometimes I have with pleasure, a great Engine, wherein the Works of I know not how many Trades, and a great many other Motions, were performed by little Puppets [. . .] and all these set a work by one Spring, which communicated Motions that were regulated and determined by the particular structure of the little Statues and other Bodies [. . .] I cannot think it impossible that the Divine and Great Δημιουργός [demiurge] [. . .] should be able by the Motions and Structures of Matter, to set a work very many Partial and Subordinate Engines.[56]

Boyle urged that empirical natural philosophy, understood both as a set of practices for investigating the world and as a mode of explanation that purported to rely wholly on sensory experience for its stock of ideas and images, could make the wisdom of God into an object of knowledge. To explain the design of natural things in mechanical terms was, he suggested, to give a uniquely intelligible account of how God imposed his intentions onto nature, whether the laws of motion given to all material things or the care taken to equip animals with sight.

The physico-theologians who published after Boyle were not convinced that all the purposes in nature could be accomplished by mechanical means alone. Grew and Ray both argued that complex forms of purposiveness, such as the instinctive ability of birds to build nests and the growth of plants and animals from seeds and eggs, required the direct supervision of another immaterial, intelligent agent. Referred to by Cudworth in the *True Intellectual System* as a "Plastick Nature," this agent was a subordinate spiritual entity that enabled nature to fulfill purposes too complex to be accounted for by mechanical explanations but not sufficiently dignified to merit God's immediate intention. It is crucial to recognize, however, that Ray and Grew regarded this agent as one that operated in concert with mechanism, guiding it in its most complex functions rather than supplanting it or making it unnecessary.[57] As a result, they could use mechanical explanations to account for the material causes of purposiveness in nature even though they believed that immaterial agents were also at work. When representing design in plant and animal bodies, indeed, the two naturalists frequently used the same strategies that Boyle had recommended in *Final Causes*. Recall that Ray, for instance, had no recourse to a "Plastick Nature" when dealing with human and animal eyes in *The Wisdom of God*. Instead, he depicted parts such as the ciliary processes and the nictitating membrane as machines that could be thought to accomplish their purposes in the same manner as devices designed by humans. As we shall see in the chapters to come, Ray and his contemporaries stuck with this approach because of their shared understanding of the relationship between experience and the formation of mental ideas. Holding, like Boyle, that the mind could only reckon with representations derived from sensory experience, they tended to regard explanations framed in material terms as the only ones that the mind could access—even though they doubted their sufficiency.

If mechanical explanations derived from sensory experience made it difficult to account for immaterial agents, they nevertheless came with crucial advantages. We have already seen that Boyle and his contemporaries positioned their empirical approach to the study of design in nature as one that could procure a far more detailed and intelligible account of God's involvement with the world than any other form of natural theology. They did not, however, see physico-theology as an exercise in the dispassionate production of knowledge. Instead, as we shall see in chapter 5, they asserted that the experience of nature's design would result in feelings of pleasure, both bodily and intellectual. Moreover, they repeatedly indicated that the affective dimensions of this encounter were integral to the work of physico-theology. In the *Christian Virtuoso*, Boyle asserted that the intensive work of finding out the exact mechanisms of final causation would lead to an affective transformation:

> And 'tis not by a slight survey, but by a diligent and skilful Scrutiny, of the Works of God, that a Man must be, by a Rational and Affective Conviction, engag'd to acknowledge with the Prophet, that the Author of Nature is *Wonderful in Counsel, and Excellent in Working*.[58]

For Boyle, therefore, the sensory investigation of natural phenomena would lead not only to the acquisition of knowledge but also to the development of an appropriate disposition toward God. Ray used a similar vocabulary to describe that disposition, referring to what Boyle had termed a "rational and affective conviction" as "Affections and Habits of Admiration, Humility and Gratitude." The physico-theologians did not make it clear whether the study of design would lead to a bodily passion provoked by the pleasure of witnessing God's handiwork or a spiritual motion resulting from rational understanding. Either way, however, they asserted that endlessly detailed representations of design in nature could serve as powerful instruments of both affective and moral transformation.[59]

Sensory Experience and Imperceptible Entities

Ray and his contemporaries positioned physico-theology as an exercise in the empirical production of theological knowledge. Seeking to surpass the forms of natural theology that had come before, they purported to explain precisely how God imposed his wisdom and intentions on the parts of creation open to sensory experience. Pursuing this end, moreover, they did not simply dust off an old design argument and point out its approximate compatibility with their empirical methods. Instead, they sought to show that their strategies for making the mechanical causes of physical phenomena into objects of knowledge could yield a highly detailed and accurate account of the workings of divine wisdom. To point out that Ray, Boyle, and others saw themselves as purveyors of theological knowledge is not, however, to deny that physico-theology was at the same time highly rhetorical. We have already seen that the physico-theologians described the experience of coming to understand nature's design as one that would not only lead to sensory pleasure but also induce a profound affective transformation. It would be a mistake, however, to dismiss these claims as a rhetorical garnish, intended for no other purpose than to strengthen the appeal of otherwise difficult arguments and complex subject matters. The following chapters will instead show that such claims about the affective dimensions of imaginative and sensory experience, along with rhetorical strategies designed to provoke the kinds of pleasure described in physico-theology, had an important place in the work of natural philosophy. They will show that the regime of affective responses and dispositions laid out in physico-theology can shed light on how natural philosophers thought the bodily organs of sensation and cognition could be made into effective instruments for obtaining knowledge about the causes and purposes of natural phenomena.

Before moving on, however, it will be necessary to consider two pressing matters, both of which also concern the place of sensory and imaginative experience in seventeenth-century strategies for learning about nature's design. The first of these is the difficulty naturalists had in securing agreement about how to identify design in natural things, or even whether everything in nature could be regarded as the product of such design. This difficulty will be the subject of chapter 3, which deals with questions and controversies raised by naturalists including Nehemiah Grew, Robert Hooke, and Edward Lhwyd about how to interpret natural things that did not always appear to have the purposiveness or beauty that they expected from the works of a divine designer. The second issue, meanwhile, concerns the dilemmas arising from the efforts made by naturalists, whether in physico-theology or natural philosophy, to somehow infer the existence and behavior of completely imperceptible entities through investigatory practices that depended wholly on sensory experience. Perhaps the most obvious example of such efforts is to be found in a case that we have already briefly considered—Ray and Grew's desire to show that the design in nature resulted not only from mechanisms imposed by God but also from the activity of intelligent, immaterial agents subordinate to God. William F. Bynum and Brian Garrett have shown that Ray and his contemporaries often used what could be called "negative induction" to prove this point, inferring immaterial agency when they identified the phenomena in question as too complex to be accounted for by mechanical causes alone. In *The Wisdom of God*, for instance, Ray asserted that no machine could produce the seeming intelligence shown by birds in feeding and housing their young. He proposed instead—in part, to avoid giving animals souls of their own—that such instinctual behaviors were compelling demonstrations of the otherwise imperceptible activity of the intelligent agents used by God to make nature accomplish complex purposes.[60]

Dror Wahrman and Jonathan Sheehan have recently pointed out that this seemingly incongruous commitment to evincing the activity of imperceptible spiritual entities by empirical means was widespread in English natural philosophy of the later seventeenth and early eighteenth centuries. Rather than marking a radical departure from a previously mechanical view of the world, indeed, they suggest that Isaac Newton's apparent willingness to admit occult causes for gravitation was congruent with earlier explanations for complex purposiveness in nature, such as Ray and Cudworth's "Plastick Nature," premised on immaterial, intelligent agents.[61] As we have seen from our reading of *Final Causes*, Boyle came much closer than either Ray or Grew to giving a fully mechanical account of purposiveness in nature. Margaret J. Osler has shown, however, that even Boyle admitted immaterial agents, or at least ones that were not straightforwardly material, at crucial points in his natural philosophy. In *The Origine of Formes and Qualities* (1666), for instance, he also expressed doubt that mechanical causes alone could account for "so orderly and well contriv'd a Fabrick as This World." Like Ray and Grew, he thus argued that intelligent agents must be at play, suggesting both that God gave additional guidance to the motions of matter in the first stages of the world's development and that God implanted so-called seminal principles into the bodies of minerals, plants, and animals to guide such complex processes as generation and reproduction.[62] As Peter Anstey has pointed out, Boyle never quite made the ontological status of those seminal principles clear, sometimes hinting that they were immaterial and on other occasions indicating that they might have mechanical causes after all. Either way, it is evident that he resorted to such unintelligible forms of agency, inaccessible to sensory experience and unimaginable as the outcome of mere mechanism, when he believed that mechanical causes alone could not explain the phenomena in question.[63]

The physico-theologians interpreted the evidence of sensory experience in two contrasting ways. On the one hand, as we have seen, they claimed that God's wisdom and purposes could be made into objects of empirical knowledge, arguing that design in nature could be understood to result from the same mechanical causes open to the senses in the world of material things. On the other hand, they also wanted to show that the agents ultimately responsible for setting the mechanisms of final causation into motion and guiding them in their most complex acts belonged to an ontological realm of spiritual entities beyond the one directly accessible to sensory experience. When making this latter point, therefore, they turned to evidence seeming to suggest instead the unintelligibility of natural processes, arguing that mechanical causes alone were incapable of explaining final causality in its most complex manifestations. In some measure, these contrasting approaches reflected the competing demands imposed on physico-theology by the polemical context—particularly the perceived threat of Epicurean materialism. Recall that Epicurus had hypothesized a world dependent on a chaotic substratum of atoms colliding at random, using this lack of order to argue that the world had no need of divine guidance. It was not enough simply to show that empirical methods could reveal the material parts of the world to depend instead on an intelligible order of mechanical causes. It was necessary to show, simultaneously, that the same methods could vouchsafe the otherwise unintelligible activity of God and his subordinate spiritual agents. Ray and his contemporaries set out to make the imperceptible world of immaterial entities into an object of empirical study—even as they denied such a thing to be possible.

In this chapter, we have seen that physico-theology depended far more on explanatory and representational practices derived from natural philosophy than has hitherto been realized. It would therefore be a mistake to dismiss its concern with verifying the existence of imperceptible entities, and thus the existence of an entire ontological order, as just one more apologetic stratagem. In the next chapter, therefore, I shall instead place such efforts to bring imperceptible entities into the realm of experience at the center of a broader reconsideration of the empirical enterprise, whether considered in its theological or its philosophical manifestations. Indeed, by examining how Boyle and others addressed concerns about the representation of such entities, we will be enabled to answer some fundamental questions. The consideration of imperceptible entities makes it possible, for instance, for us to examine how the physico-theologians addressed the limits of the senses, drawing on contemporary theories of sensation and cognition in the process. Such theories were invariably freighted with theological significance, reproducing the same ontological questions that troubled physico-theologians in their efforts to work out which parts of the mind depended on bodily mechanisms and which ones depended instead on the immaterial soul. Inevitably, therefore, attempts to pin down the limits of sensory experience brought up questions about the extent to which the human mind had the capacity to settle metaphysical and theological questions—questions, that is, about the fundamental conditions of existence and God's role in establishing them. By reckoning with the problem of imperceptible entities, therefore, we will be able to reconsider long-running questions about the relationships between theology, metaphysics, and natural philosophy in the second half of the seventeenth century.

2

An Empiricism of Imperceptible Entities

Physico-theology purported, as we have seen, to make the attributes of an immaterial God known by empirical means. The difficulty with such a claim, however, was that such a God was thought to lie far beyond the grasp of the senses. All but the most heterodox thinkers of the seventeenth century agreed that God was an entity both infinite and composed of an immaterial substance. The most widespread and influential contemporary accounts of the mind's workings held, moreover, that such entities were simply incapable—at least in nature's regular course—of affecting the bodily organs involved with sensation and the subsequent presentation of ideas to the mind. An incorporeal substance could never make impressions on the corporeal organs of sensation, while an infinitely extended entity could never be represented as an object of experience. In his *Discourse of Things Above Reason* (1681), Robert Boyle therefore warned that when attempting to contemplate God's infinite perfections, it was "no less weakness than presumption to imagine that such finite Beings as our Souls, can frame full and adequate *Idea*'s of them." The problem, then, was that Boyle and the other physico-theologians wanted to use empirical means to learn about a God who was widely understood to be both imperceptible and completely inconceivable.[1]

It was not only the spirit world, however, that posed an acute challenge to the empirical project. There were also many objects belonging to the world of material things that were equally difficult to make accessible to the bodily organs of sensation and cognition. Boyle's famous programs of pneumatic and chemical experimentation are a case in point. Boyle used those experiments to establish the famous corpuscular hypothesis, which attributed the properties perceptible in material things to mechanically configured interactions between imperceptible clusters of infinitesimally small atoms of matter. For the most part, Boyle called those semipermanent clusters "corpuscles." Meanwhile, he generally referred to the even smaller fundamental particles out of which those corpuscles were composed as "atoms," *minima naturalia*, or *prima naturalia*. Boyle was not entirely consistent, however, sometimes calling those fundamental particles "corpuscles" as well. When appropriate, therefore, I will follow William R. Newman in referring to the semipermanent clusters on which Boyle's theory of matter mostly depended as "aggregate corpuscles" to distinguish them clearly from their atomic constituents.[2]

Even though Boyle regarded the aggregate corpuscles and their fundamental constituents as crucial to his account of physical causation, he frequently admitted their inaccessibility to the human faculties. This was a point that he also made in *Things Above Reason*, explaining that the imagination—the faculty responsible for presenting ideas of things to the mind—was incapable of picturing particles so small and so great in number:

> So if you would imagine an Atome, of which perhaps ten thousand would scarce make up the bulk of one of the light particles of dust, that seem to play in the Sunbeams when they are shot into a darkened place, so extraordinary a littleness not having fallen under any of our Senses, cannot truly be represented in our imagination.[3]

In *Things Above Reason*, Boyle characterized both the immaterial God and the imperceptible particles God had supposedly created as inaccessible to sensory experience. Boyle indicated, if only indirectly at this point, that practitioners of physico-theology and natural philosophy alike had to answer the same fundamental question: Could one gain access to entities that were both completely imperceptible and utterly inconceivable through encounters with things capable of affecting the bodily organs involved with sensation and cognition?

Considering the problem of imperceptible entities, we can therefore answer fundamental questions about what philosophers of the seventeenth century thought it meant to seek out knowledge of the world by empirical means. As we saw in the last chapter, thinkers such as Boyle and John Ray closely linked the challenge of showing that the world depended on orderly interactions between imperceptible particles of matter to the task of demonstrating the existence and activity of an immaterial God. In this chapter, meanwhile, we shall see that philosophers who wanted to reckon with both those imperceptible entities also had to reckon with more prosaic problems concerning the limits of the sensory and cognitive organs and the extent to which it might be possible to overcome those limits. It only takes a glance at contemporary works of neurophysiology to see that questions about how the brain and senses worked were inextricably tied up with questions of metaphysics and ontology—questions about the existence and workings of the world's material and immaterial constituents. For philosophers such as René Descartes and Thomas Willis, the aim of neurophysiology was not merely to understand the mechanisms at play in the mind's bodily constituents but also to verify the existence of an immaterial soul and its role in the mind's higher functions. Inevitably, therefore, considering the limits of the senses required philosophers to pose theologically freighted questions about the possibility of verifying the existence and operations of otherwise imperceptible entities. Whether answered affirmatively or negatively, questions about whether the mind could gain access to imperceptible entities were also questions about how the parts of the order of being, from imperceptible particles of matter to immaterial spirits, were related to one another. Our aim will therefore be to do more than simply piece together the strategies that Boyle and his contemporaries used to make imperceptible entities accessible to the senses. We will at the same time work out some of the ways in which those strategies were tied up with ideas about God's relationship to the world, both to the humans and to the rest of creation.

For much of its length, this chapter will focus on two books by Robert Boyle. The first will be the aforementioned *Discourse of Things Above Reason*, a dialogue in which Boyle sought to describe the limits of the human mind in relation to both theological and philosophical problems. The second, *Some Physico-Theological Considerations about the Possibility of the Resurrection* (1675), attempts to consider the same issues in relation to a single question: whether the resurrection promised by the Bible to humanity at the end of time was physically possible. With an experimental demonstration of the corpuscular hypothesis as its centerpiece, this short but audacious text gives us a remarkable opportunity to see how Boyle mobilized strategies for representing imperceptible atoms in his efforts to establish what the human mind could learn about God. With this focus on Boyle, moreover, we can reconsider the relationship between his approach to the production of physico-theological knowledge and the approaches pursued by other members of the Royal Society of London, such as Grew and Ray. It is well known that Boyle's theological beliefs—in particular, his beliefs about the how God imposed his will on creation—marked him out from his contemporaries. As we saw in chapter 1, however, it is not always productive to assume that theology in the seventeenth century was the preserve of belief, or even that statements of religious belief are infallible guides to the religious "content" of systems of natural philosophy that aimed to settle metaphysical questions. In this chapter, therefore, we will compare Boyle's statements of belief with the strategies that he used for making both God and atoms accessible to the sensory and cognitive organs. In doing so, we shall see that Boyle's ontological assumptions were sometimes closer to those of his contemporaries than has often been thought. Like Ray and Nehemiah Grew, Boyle was caught between two almost contradictory desires, hoping to make God into an article of empirical knowledge while at the same time seeking to show that God was thoroughly inaccessible to the human faculties.

Imagination, Intellect, and an Inconceivable God

Perhaps the deepest paradox of the empirical enterprise was that Boyle and his contemporaries subscribed to a model of the human faculties that made it impossible for the mind to picture things that lay beyond the realm of sensory experience. As articulated, for instance, by Descartes in the *Treatise on Man* (posthumously published in 1662) and *Meditations*, this model depended on a distinction between the functions of the imagination and the intellect. The imagination, Descartes argued, was a bodily organ responsible for presenting the mind with representations of things derived from sensory experience. The intellect, meanwhile, was a faculty of the immaterial soul, capable not only of subjecting those representations to reason but also of intuiting the existence and attributes of entities that could never affect the senses. For Descartes, therefore, the only way to verify the existence of entities that lay beyond the grasp of the senses—whether because they were extremely small or because they were immaterial—was by turning to the immaterial intellect. It was simply impossible, as he saw it, for the bodily organs of sensation and cognition to present the mind with images of things that they could not perceive.[4] The strange thing, as we will see, is that empiricists such as Boyle, Grew, and Ray broadly concurred with this scheme of the mental faculties and even registered their agreement with the notion that the senses were as good as useless when it came to imperceptible entities. At the same time, however, they also insisted that the best way of giving the mind access to imperceptible entities was by empirical means—through the very organs that seemed incapable of detecting them. Unlike Descartes and other rationalist philosophers, they made the mind dependent for crucial insights on the bodily organs involved with sensory perception. To understand how Boyle and his contemporaries sought to learn about imperceptible entities, we must therefore begin by considering the embodied mechanisms of sensation and cognition upon which they thought the mind depended.

For philosophers of the seventeenth century, the task of explaining how the brain and nervous system worked was inextricably bound up with the task of precisely delineating which of the mind's functions depended on the agency of those bodily organs and which depended instead on the agency of the immaterial soul. This question was, for example, central to the work of Thomas Willis, the Oxford physician and fellow of the Royal Society responsible for the two most influential works of neurophysiology published in England during the seventeenth century. In the first of these, *Cerebri Anatome, cui accessit Nervorum Descriptio et Usus* (1664), he offered an extremely detailed description of the brain and nervous system based on human and animal dissections, accompanied by an ostensibly empirical account of how those organs worked.[5] In the second, *De Anima Brutorum quae Homine Vitalis ac Sensitiva est* (1672), he used observations on the nervous systems of animals, along with pathological observations on nervous distempers in humans, both to delineate the bodily causes of nervous, affective, and cognitive functions and to distinguish them from the operations of the incorporeal soul.[6] There was no question of making the soul itself visible, so Willis sought instead to demarcate its functions by comparing the structures of the brain and nervous system in humans and animals. Since animals were thought not to possess souls, he could use the similarities between their anatomy and that of humans to isolate the functions brought about through the physical agency of bodily organs from those caused by the imperceptible influence of the incorporeal soul. He therefore attributed many of the functions of the brain and nervous system, from the performance of involuntary functions to simple forms of cognition, to the workings of an assemblage of bodily mechanisms shared by humans and animals alike. Willis referred to this assemblage as the "animal soul." Meanwhile, he could assign the higher functions seemingly possessed by humans alone, such as the capacity to make moral judgments or ponder abstract ideas, to the unique possession of what he called the "rational soul."[7]

Crucially, Willis included the imagination and memory among the faculties shared by humans and animals and therefore attributable to the bodily agency of the brain and nervous system.[8] He hypothesized that the formations of ideas began when the impressions made by external things on the sense organs were transmitted through the nerves toward the brain by undulating waves of spirituous particles that he called the "animal spirits." Those impressions had to be made forcefully, however, if they were to result in either the formation of mental images or their retention in the memory. Weaker impressions, Willis supposed, would only reach as far as the cerebellum, provoking merely the involuntary responses for which he held that organ responsible. If the impressions were strong enough, however, they would travel farther, passing through a body at the base of the brain called the *corpus callosum* to reach the *cerebrum*. It was in the corpus callosum, Willis argued, that ideas were generated, taking form as imperceptibly small material impressions made on that body. Finally, the most forceful impressions would pass from there to the *cerebral cortex*, where the ideas generated in the cerebrum could be laid up for future use as memories.[9] For Willis, the generation of ideas was a mechanical process, dependent on the force with which external things acted on the organs of sensation, imagination, and memory.

We can get a sense of how literally Willis meant this embodied, mechanical account of ideation by noting that he even saw the comparison of ideas as a material process, picturing that crucial cognitive act as a collision between tiny material entities passing through the brain. Indeed, he even used differences in the form of the cerebrum in various animals to explain differences in their imaginative abilities. He pointed out, for instance, that the cerebrum in cats had a simple and regular structure, its few twists and turns providing little occasion for ideas to meet in new or unexpected combinations. Such a structure, he suggested, was ideally suited to the limited imaginative requirements of a creature that only ever had to follow its instincts. The human cerebrum, meanwhile, was far more complex, characterized not only by a greater number of folds but also by a seemingly random order in their disposition. The convoluted form of the human cerebrum permitted the mind to range ideas in a far wider range of combinations, facilitating the higher cognitive functions that distinguished humans from beasts:

> Hence these folds or rollings about [in the cerebrum] are far more and greater in a man than in any other living Creature, to wit, for the various and manifold actings of the superior Faculties; but they are garnished with an uncertain, and as it were fortuitous series, that the exercises of the animal Function [the imagination] might be free and changeable, and not determined to one.[10]

If the human cerebrum was more complex than that of the cat, it was nevertheless explicable in the same terms. For Willis, the formation and comparison of ideas was a bodily process, resulting ultimately from the motions induced by external things when they touched the sensory organs.[11]

Willis's hypothesis provoked widespread disagreement, with commentators pointing out the lack of empirical evidence for the imperceptible animal spirits so crucial to his explanation for the workings of the nervous system.[12] For Grew and Ray, the main difficulty was supplied by their doubt—familiar to us from chapter 1—that so many complex nervous and cognitive functions could be attributed to the blind mechanical agency of material things.[13] In his *Cosmologia Sacra*, for instance, Grew argued that the imagination required immaterial supervision, making his case by pointing out the imagination's well-known capacity to improve upon experience by placing ideas in combinations that had never been witnessed before:

> The Power of Phancy [imagination] appears, not only in Drawing the Schemes of Things according to Sense: But also in such a manner, as they never were in Sense: As of a Horse, eating a Lobster. The parts indeed, of this Image, the Horse, and the Lobster, are both derived of Sense. But this Composition, never was in Sense, nor ever will be, but only in Conceit.[14]

Although Grew's argument here was at odds with Willis's more materialist account of the imagination, his words nevertheless reveal a crucial point of agreement. While seeking to show that only an immaterial agent could put ideas into new combinations, he nevertheless admitted the only ones to which that agent had access were those derived from sensory experience. At other points, meanwhile, he confirmed Willis's suggestion that ideas were in fact material entities lodged in the brain. He described the gradual fading of memories, for example, as a bodily process in which "the Impressions are obliterated, or the Images dissolved into their first Principles, or Exterminated from the Brain, with the Current of the Animal Spirits into the Nerves."[15] While disputing the exact division of the mind's functions between its material and immaterial constituents, the two philosophers—along with contemporaries such as Boyle—nevertheless agreed that the imagination depended wholly on the bodily mechanisms of the sense organs for its stock of ideas.[16]

For much of the seventeenth century, attempts to settle the division of mental functions between matter and spirit were directly relevant to enormously important debates about whether, or under what conditions, God could be made into an object of knowledge. In his *Leviathan* (1655), for instance, Hobbes diagnosed competing claims to knowledge of God as one of the leading causes of political strife. In pursuit of a lasting political settlement, therefore, he sought out a way of excluding religion from the domain of reason, making it a matter of private belief rather than public knowledge.[17] Hobbes worked toward this end by arguing that the mind was a wholly material entity, thereby making it completely dependent on the senses and imagination for its supply of ideas. With this model of the mind, he could attack what he took to be the epistemological pretensions of theology simply by pointing out that the imagination was incapable of representing the immaterial spirits whose existence it sought to demonstrate:

> But the opinion that such Spirits were Incorporeall, or Immateriall, could never enter into the mind of any man by nature; because, though men may put together words of contradictory signification, as Spirit, and Incorporeall;

> *yet they can never have the imagination of any thing answering to them*: And therefore, men that by their own meditation, arrive to the acknowledgement of one Infinite, Omnipotent, and Eternall God, choose rather to confesse he is Incomprehensible [. . .] than to define his Nature By Spirit Incorporeall, and then Confesse their definition to be unintelligible.[18]

The consequences of Hobbes's account of the mind were therefore far-reaching. His argument that immaterial spirits were inaccessible to a material mind made them quite literally unthinkable. Since the mind was incapable of representing immaterial things to itself, he argued, it was fruitless to try to reason about them.

For Boyle and the other proponents of physico-theology, all of whom wanted by contrast to show that it was possible to obtain knowledge of immaterial agents, this position was unacceptable. Since they agreed with Hobbes's account of the imagination, however, they did nothing to challenge his argument that those agents were unimaginable. Instead, they argued that the immaterial part of the mind could intuit things that the its bodily organs would otherwise be unable to grasp. In *De Anima Brutorum*, for example, Willis argued that the "rational soul" could access ideas that were "by no means to be learnt from Sense or Phantasie [imagination]." Among these, he included some of the most important tenets of natural and revealed religion, including God's infinity, the requirement to worship God, the existence of incorporeal beings such as angels, and the existence of heaven and hell.[19] In *Things Above Reason*, meanwhile, Boyle used the same reasoning to criticize Hobbes's argument—or at least an argument very much like it—that immaterial entities were inconceivable because they could not be imagined:

> Those that are wont to employ their imaginations about things that are the proper Objects of the Intellect, are apt to pronounce things to be unconceivable, only because they find them unimaginable; as if the Fancy and the Intellect were Faculties of the same extent: Upon which account some have so grossly err'd, as to deny all immaterial Substances.[20]

Boyle identified Hobbes's error as a failure to recognize the difference between the imagination and the intellect. He agreed that any attempt to represent immaterial entities with the imagination would end in failure. He did not agree, however, that the imagination was the mind's only source of ideas. Echoing Descartes's argument in the *Meditations*, Boyle claimed instead that the possession of an immaterial intellect enabled humans to reason about incorporeal beings even though they would be forever inaccessible to the senses and imagination.[21]

A glance back at their physico-theological works reminds us, however, that Boyle, Grew, Ray, and Willis were far more interested in using the organs of sensory perception and imagination to learn about God than they admitted in this context. As we saw in chapter 1, they claimed that the empirical study of nature could demonstrate the existence and activity of God, and even gain a measure of insight into God's attributes. In his *Disquisition about the Final Causes of Natural Things*, moreover, Boyle criticized Descartes's approach to proving God's existence—based solely on intuitions performed by the intellect without the benefit of experience—for revealing too little of God's wisdom and goodness. Boyle and his contemporaries therefore held positions that seem contradictory. When answering the challenge posed by materialist accounts of the mind, they emphasized the impossibility of subjecting an immaterial, infinite God to the bodily mechanisms of sensory experience. In the context of physico-theology, however, they asserted that sensory experience was a highly effective means of giving the mind access to such a God. The premise of physico-theology was therefore precarious and easy to contest. It stood the risk of being undermined by an account of the mental faculties with which Boyle and his contemporaries evidently agreed, at least in some contexts. We are left to wonder how Boyle and his contemporaries could reconcile such apparently contrasting claims.

How to Represent an Unimaginable God

The *Discourse of Things Above Reason* was Boyle's most highly developed reflection on the question of God's accessibility to the human faculties. Examining this work in greater detail will enable us to see how he attempted to balance the assertion of God's unintelligibility against a desire to use the imagination and senses as instruments of theological knowledge. Despite the work's presentation as a friendly dialogue among four participants, its argument appears at first glance to be unambiguous. For Boyle, some matters lay completely beyond the reach of the human mind, inaccessible both to its bodily organs and to the immaterial but finite human soul. He described such matters as "things above reason." This category included points of Christian doctrine that defied rational explanation, such as the apparently irreconcilable coexistence of God's necessary capacity to know the outcome of every future event alongside the human lived experience of free will, and the apparent importance of individual agency in the Christian economy of future rewards and punishments.[22]

Boyle's ostensible reason for writing *Things Above Reason* was to persuade philosophically minded readers not to reject Christianity because it contained doctrines that defied rational explanation or even comprehension. In line with this agenda, he made his case by seeking to remind those readers how often the pursuit of mathematics philosophy involved accepting similarly unintelligible notions. Among the most salient of those were hypotheses about the forms and behavior of imperceptible atoms and the consideration of infinite quantities in mathematics and natural philosophy. Indeed, he argued that the paradoxes flowing from the contemplation of infinites could serve to illustrate the impossibility of fully understanding the nature of God. There was therefore no reason, Boyle insisted, to regard matters of religious belief as the only ones involving propositions that transcended the capacities of reason. Rather, he suggested that both philosophers and theologians sometimes found themselves in situations that required them to accept as truths things impossible to reconcile with human reason. Boyle divided those truths into three kinds: "incomprehensible," "inexplicable," and "unsociable." The infinite perfections of God were *incomprehensible* because the finite human faculties could never form adequate ideas of them. Meanwhile, *inexplicable* truths were those that could be perceived or intuited but not explained. Boyle illustrated this kind of truth using the paradox of the infinite divisibility of matter, a point that could be demonstrated mathematically but that defied physical explanation. Finally, *unsociable* truths—and here Boyle had in mind the difficulty of reconciling divine foreknowledge with the experience of free will—were those that appeared to contradict one another. He exemplified this final category by referring again to infinite divisibility, pointing out the apparent contradiction involved in the mathematical demonstration that lines of different lengths must nevertheless be divisible into the same infinite number of parts.[23]

Since Boyle generally adopted a conciliatory authorial persona, addressing debates indirectly and rarely naming his opponents, it can be hard to identify the targets of his apologetic and controversial works. In *Robert Boyle and the Limits of Reason* (1997), however, Jan W. Wojcik argues that Boyle's underlying aim was to intervene in two contemporary theological debates. The first, which had been brewing in England since the 1650s, concerned the perceived threat of Socinianism, a brand of theology that argued that the truth of scriptural revelation ought to be judged according to the strictures of human reason. Socinians thus rejected important Christian mysteries, such as the resurrection of the dead and the Last Judgment, because they seemed impossible when subjected to rational scrutiny. To neutralize this attack on revealed religion, Boyle sought instead to show that those mysteries exceeded the grasp of reason and that it was thus inappropriate to rely on reason alone when deciding whether they were true or false. It was futile, he suggested, to make reason the sole judge of things that far transcended its capacities.

At the same time, Wojcik continues, Boyle wanted to address a recent outbreak of the interminable debate about predestination. For Calvinists, God's necessary foreknowledge of future events implied that free will was more or less illusory—individuals were predestined for heaven or hell by divine decree. From Anglicans to Nonconformists, however, a wide range of contemporaries found predestination inconsistent with God's goodness because it seemed to make God responsible for human sin. On this point, Boyle saw an opportunity to calm disputes between rival groups of fellow Protestants, arguing that predestination was one of those matters that lay beyond the grasp of reason. With this maneuver, he hinted that nobody could ever hope to solve the paradoxes of predestination by rational means, and thus also hinted at the futility of conflict among Christians on such thorny points of doctrine. Neither side, in Boyle's view, could legitimately claim a greater share of truth than the other.[24]

At one level, therefore, Wojcik explains Boyle's position in terms of his involvement with contemporary theological controversies. She goes on, however, to propose that there was an even more powerful agent at work—Boyle's religious belief. Indeed, she identifies this belief as the motivation not just for his theology but also for his natural philosophy, writing that "Boyle's theological beliefs provided the foundation for his views on natural philosophy."[25] It is well known that Boyle was committed to a position now referred to as *theological voluntarism*, a position that is usually contrasted—sometimes too sharply—with another view called *intellectualism*. For an intellectualist such as Henry More, the world could be understood through systems of reasoning such as logic and mathematics because it was a stable product of divine wisdom, with God in some sense constrained by the logic of his own decrees. For voluntarists, on the other hand, the world was utterly contingent on arbitrary decisions emanating from God's free will. Boyle's God could therefore institute laws of nature but nevertheless choose to suspend those laws if, for instance, he wanted to communicate with humans through a supernatural sign. Scholars such as Margaret J. Osler have shown that theological voluntarism played a decisive role in the emergence of the empirical philosophies of Boyle and Pierre Gassendi, the French thinker who did more than just about anybody else to bring Epicurean atomism into the European intellectual mainstream. Since they regarded the world as more a product of arbitrary will than rational order, Boyle and Gassendi saw little point in mimicking Descartes's attempt to force natural philosophy into a logically coherent system of reasoning. Instead, they sought to learn about the world by gathering histories and other testimonies of nature in action. Osler therefore regards Boyle's epistemology—which sought only probability rather than demonstrative certainty—as the product of his theological voluntarism.[26]

To this familiar account, however, Wojcik adds something more controversial. She claims that Boyle, extending the logic of voluntarism to his account of the human faculties, believed human and divine reason to operate according to entirely different principles. Boyle's belief in theological voluntarism led him, on this account, to put human reason and divine wisdom into completely different categories, effectively making God wholly inaccessible to rational investigation.[27] She contends, moreover, that Boyle's famous commitment to probabilism was equally symptomatic of this deep-seated irrationalism. She argues that Boyle limited his natural philosophy to the production of probability rather than demonstrative certainty because he saw both the world and the human faculties as contingent expressions of God's arbitrary will.[28]

It is clear from Boyle's stated position in *Things Above Reason* that he sometimes held and expressed such a belief in God's inaccessibility to reason. It is less clear, however, that this belief decisively shaped his ideas about the conditions under which knowledge of imperceptible entities—whether of atoms or of God—could be obtained. When, for instance, he turned to examples from natural philosophy to illustrate the mind's encounter with God, the resulting claims were extremely ambiguous. Consider a passage from the second half of *Things Above Reason* where Boyle, for the second time, compared the intellect's inability to grasp God's attributes with the imagination's inability to picture countless infinitesimally small atoms:

> I see no necessity, That Intelligibility to a *humane Understanding*, should be necessary to the Truth or Existence of a thing; any more then that Visibility to a *Humane* Eye, should be necessary to the Existence of an Atome, or of a Corpuscle of Air, or of the *Effluvium's* of a Loadstone, or the Fragrant Exhalations of *Ambergris*, and Musk from a perfumed Glove.[29]

As we saw at the very start of this chapter, Boyle and his contemporaries frequently discussed the impossibility of experiencing or imagining imperceptibly small particles of matter. Indeed, they held up the imagination's inability to picture the enormous number of atoms making up even the smallest perceptible things as an appropriate analogue for the intellect's inability to grasp God's infinite perfections. In his *Cosmologia Sacra* (1701), for example, Grew used the inconceivable divisibility of matter, infinite for all practical purposes, as an illustration of how "deeply and far out of sight" God had hidden "the Foundation of the Generations and Operations of Bodies."[30] Boyle's comparison between atoms and God might therefore be cited as an expression of his radical voluntarism, with its emphasis on the impossibility of perceiving atoms interpreted as a reminder of the discrepancy between human reason and God's will.[31]

Given a page later, however, Boyle's own explanation gives us quite a different sense of his intentions. He clarified his meaning by mobilizing another comparison from the domain of philosophy, this time involving geometry:

> There is no necessity that we should be interdicted all Knowledge of those sublime Objects, in which there are many things, whereof [. . .] we must confess our selves ignorant. Thus elder Geometricians knew very well what a Rectangular Triangle was, when they conceived it to be a Figure consisting of three strait Lines, two of which comprize a right Angle; though probably for a great while they did not know so much as all its *chief Properties* or Affections.[32]

Geometricians before Pythagoras knew exactly what a right-angled triangle was—a triangle with a ninety-degree angle in one of its corners—even though they had not yet worked out that the square of the hypotenuse was equal to the sum of the squares of the remaining two sides. For Boyle, therefore, the history of geometry showed that accurate and useful knowledge of an object did not have to depend on a complete intellectual or imaginative grasp of all its properties. Extending this line of thought to the "sublime Objects" of theology, Boyle suggested that knowledge of God was possible after all. The human faculties could obtain partial but nonetheless informative knowledge of God's attributes, even though it would never grasp them entirely.[33]

Boyle gave several indications, moreover, that he thought it possible to obtain knowledge of God's existence and attributes through the contemplation of material things capable of affecting the bodily organs involved with sensation and imagination. Consider, for instance, his treatment of a thought experiment from the sixth of Descartes's *Meditations*. Descartes had begun by pointing out that it was easy both to understand and to form a mental image of a three-sided figure: "not only do I understand it to be a shape enclosed by three lines, but at the same time, with the eye of the mind, I contemplate the three lines as present, and this is what I call imagining." The same was not true, however, when the mind attempted to reckon with shapes possessing many more sides. It was easy enough to understand that a "chiliogon" was a figure with a thousand sides. It was evidently impossible, however, to form an accurate mental image of such a complex shape. "I may perhaps picture some figure to myself in a confused fashion," Descartes explained, "[but] it is quite clear that this is not a chiliogon, because it is not at all different from the picture I would also form [. . .] if I were thinking about a myriogon, or some other many-sided figure." In other words, the intellect could understand complex phenomena even when the imagination was incapable of picturing them. For Descartes, the contemplation of many-sided polygons served to demonstrate that the intellect was an immaterial entity that did not depend on the bodily faculties of the mind—the senses and the imagination—for its existence.[34]

We saw earlier that Boyle used the same ontological distinction at another point in *Things Above Reason* to attack Hobbes's argument for the impossibility of reckoning with immaterial entities. Here, however, he glossed over that crucial distinction, instead making an argument that identified similarities between the two faculties. He proposed to explain the immaterial intellect's encounter with God by likening it to the imagination's encounters with material things:

> 'Tis scarce possible to find very apposite examples, to illustrate things of a kind so abstruse and heteroclite as those may well be suppos'd, that do surpass our Reason. But yet some assistance may be borrowed from what we may observe in that other faculty of the mind, which is most of kin to the Intellect, I mean the *Imagination*.[35]

Ignoring the differences between the intellect and the imagination enabled Boyle to put Descartes's demonstration of the weaknesses of the imagination to a new purpose. He agreed that the imagination stood no chance of forming a complete, intelligible image of a figure like the "myriogon." The result would be nothing more than "a confused *Idea* of a *Polygon* with a very great many sides." Unlike Descartes, however, Boyle asserted that the imagination's inability to picture such an object could serve as an informative stand-in for how the intellect was overpowered in its reckoning with God.[36]

Boyle would use similar comparisons to illustrate the mind's encounter with God throughout the work. In several places, for instance, he likened it to the eye's inability to grasp objects too large to be seen in one view. It was impossible, he noted in one of those comparisons, for the unaided eye to perceive the full extent of an ocean. As Boyle explained, however, no reasonable observer would count the eye's weakness as a good reason for denying that the rest of the ocean existed and that it was truly a vast object. Instead, he pointed out that a sensible observer would realize that what he could see was merely a small part of something much larger—an enormous body of water "which may, for ought he knows, reach to a vast extent beyond the visible Horizon." Elsewhere, he made the same point using images of partial vision—for instance, likening the intellect's encounter with God to a mariner catching sight of land for the first time. He might not know exactly what type of land he would find when he got closer to the shore—whether it would turn out to be rocky or inhabited—but he could nevertheless be satisfied with the partial insight that it was land and not sea. As we can see, the import of these comparisons was mixed. On the one hand, they held up the imagination's inability to reckon with extremely large or distant objects as an illustration of the intellect's inability to reason about an infinite God. On the other hand, the comparisons undercut this message by signaling that the imagination's encounters with some objects could serve as a model for grasping, if only in a partial sense, what God's attributes were like. The repeated appearance of such comparisons gives us a sense, therefore, that Boyle regarded the imagination and its objects in the world of perceptible, material finite things as informative analogues for the immaterial intellect's encounter with an infinite God.[37]

We cannot therefore regard *Things Above Reason* as a straightforward expression of Boyle's belief in God's complete inaccessibility to the human faculties. Boyle's attempts to represent the mind's encounters with God instead raised a contradictory possibility—that the imagination could enable the intellect to obtain useful knowledge of God's nature and attributes. At times this possibility, latent in Boyle's comparisons, floated to the surface of his diffusely expressed argument. Consider, for instance, his summing up of the lessons to be drawn from the comparison between the imagination and the intellect:

> So when we speak of Gods Primity (if I may so call it)[,] Omnipotence, and some other of his infinite Attributes and Perfections, we have some conceptions of the things we speak of, but may very well discern them to be but inadequate ones.[38]

Boyle did not claim here that the human faculties were totally incapable of reckoning with God. Rather, he qualified the argument he had been making up to that point with the proviso that the mind could possess "some conceptions," however limited, of things above reason. He added this qualification, moreover, at several other points in the work. When discussing eternity, for example, he remarked that "a considering person" could obtain a limited notion of infinite duration by contemplating extremely long—but finite—periods of time. In these moments, Boyle carefully distinguished his argument for the mind's inability to possess adequate ideas of an infinite God from the harsher alternative that it could possess no ideas at all. Instead of following Hobbes's assertion that the imagination's encounter with God would result in complete and utter confusion, he suggested that the senses and imagination could somehow furnish the mind with ideas bearing some relation—albeit very remote—to the attributes of God.[39]

We can now understand Boyle's position. His main aim in *Things Above Reason* was to resist the Socinian argument that revealed religion should be judged according to the strictures of human reason. For much of the work, he therefore sought to place God safely beyond the grasp of rational comprehension, mobilizing the argument that there could be no relation whatsoever between infinites and the finite objects of sensory (or even intellectual) experience. He subverted that argument, however, with his comparisons. Those comparisons brought up an opposing possibility—that the finite objects to be found among the things that God had created could, after all, yield insights commensurate in some degree to the attributes of an infinite, immaterial God.[40] An explanation for the tension between these two positions might be found by putting Boyle's desire to secure sacred mysteries from the prying eyes of reason in a broader context. In *Final Causes*, as we have already seen, Boyle set out to resist the implication that natural philosophy could do nothing at all for theology. He did not want to concede Descartes's argument that experiments and observations could reveal nothing at all of God's attributes and purposes. Even in his most concerted effort to place God beyond the grasp of reason, therefore, Boyle held on to the notion that God could be represented in terms derived from sensory experience, if only to a limited extent. In doing so, he kept open the possibility of using the practices of empiricism to represent and thereby learn about an unimaginable God.

A Chemical Resurrection

Boyle's attempt to keep open a space for physico-theology was perhaps inconsistent with his belief in a God who imposed his will arbitrarily on the world. It was entirely consistent, however, with his approach to the production of knowledge about imperceptible entities. We have already encountered evidence to suggest that he and his contemporaries identified important overlaps between the task of representing atoms and that of representing God. Recall that both Boyle and Grew held up the imagination's encounter with an almost infinitely large number of imperceptibly small atoms as an instructive analogue for the mind's encounter with God. It therefore makes sense to embark on a deeper inquiry into the possible connections between their strategies for bringing atoms into the realm of sensory experience and their attempts to do just the same thing with God.

It is true that in *Things Above Reason* Boyle used the impossibility of imagining atoms to illustrate the impossibility of completely grasping the attributes of God. His point, as we have seen, was that natural philosophers willing to accept the existence of inconceivable atoms should not turn up their noses at the possibility of an equally inconceivable God. It is abundantly clear, however, that he and his contemporaries did not admit the unintelligibility of atoms to exclude them from the realm of rational inquiry. Instead—as is well known—they claimed that observations on the objects of sensory experience could yield a partial but nonetheless highly probable kind of knowledge about the forms and motions of those particles. Boyle and his contemporaries were, moreover, aware of the resulting paradox. As they understood it, the aim of the empirical project was to drag entities that could be neither perceived nor imagined into the realm of sensory experience. In the first of his *Three Physico-Theological Discourses* (1693), John Ray was thus moved to praise Boyle's chemical experiments for their strange ability to do just that. "But for a sensible [perceptible] demonstration of the unconceivable, I had almost said infinite, divisibility of matter," he wrote, "I might refer the Reader to the Honourable Mr. *Boyl* of famous memory his Discourse concerning the strange subtlety of *effluviums*."[41] Here we can see a broad point of coherence between the natural-philosophical project of representing atoms and the physico-theological project of representing God. When Boyle likened God's infinite extent to finite objects such as the ocean or the coastline, he was marshaling the same basic assertion at work in natural philosophy: that objects of sensory experience could somehow stand in for entities that were, by their very definition, impossible to experience. To borrow Ray's magnificent turn of phrase, we might therefore say that Boyle was attempting to give his readers a "sensible demonstration" of God's "unconceivable" attributes.

In other of his works, both theological and natural-philosophical, Boyle made much more explicit claims about the theological insights to be gained from representations of imperceptible particles. Perhaps the most striking and least studied of them all is *Some Physico-Theological Considerations about the Possibility of the Resurrection*. This short text appeared as an addition to *Some Considerations about the Reconcileableness of Reason and Religion* (1675), a much longer treatise floating many of the same arguments that would receive a fuller airing some six years later *Things Above Reason*. As we saw in chapter 1, most works of physico-theology concerned themselves exclusively with natural religion, seeking only to demonstrate God's existence and activity in the world. By contrast, Boyle's *Possibility of the Resurrection* deals explicitly with a point of revealed religion—the promised recomposition of every human body at the end of time. The work therefore fell outside the mainstream, belonging to that speculative branch of physico-theology, exemplified by Ray's *Three Physico-Theological Discourses*, concerned with providential events that seemed to depend, at least in part, on physical causes—events such as the Deluge and the earth's promised destruction at the end of time. Although the work was thus a little unusual, Boyle's polemical agenda was, as Salvatore Ricciardo has recently shown, a familiar one. Boyle's polemical targets in *Possibility of the Resurrection* are the same as the ones he would go on to tackle in *Things Above Reason*. He wanted to resist the notion that revelation could be judged by exclusively rational means while at the same time demonstrating the theological utility of his own brand of atomistic natural philosophy.[42]

In line with the first of these aims, Boyle framed his discussion as a response to those who rejected the promise of the resurrection because they regarded it as impossible. For Boyle, the strongest such objection depended on the simple fact that human bodies decompose after death, their matter being widely dispersed and finally incorporated into the bodies of other substances, plants and animals. As a result, God was left with a seemingly insurmountable difficulty. To recompose every human body that had ever lived, he would first need to extricate the immense number of "scattered parts" that had once made them up from all the new substances to which they had been joined. After that, he still had the task of reconstituting all those particles "after the same manner wherein they existed" before.[43] With the title *Possibility of the Resurrection*, Boyle made it clear that he would respond to this objection neither by seeking to describe exactly how the resurrection could take place nor by attempting to prove that it could be explained in entirely natural terms. Instead, he carefully signaled his belief that so momentous an occurrence could only be regarded as an unintelligible effect of God's absolute and unfettered wisdom, will, and power. Admitting the ultimate incomprehensibility of the resurrection did not, however, lead him to make it a matter for faith alone. Instead, he argued that a limited kind of knowledge could still be obtained. That knowledge not enough to explain how the resurrection would come about, but it was sufficient at least to disabuse those who rejected the doctrine because they regarded it as a complete impossibility. If natural philosophy could not explain the resurrection in detail, it could nevertheless produce a demonstration that the process was physically possible.[44]

Here Boyle saw an opportunity mobilize the explanatory resources of his corpuscular hypothesis—the claim that transformations in the perceptible qualities of material things were caused by changes in the composition and motions of imperceptible aggregate corpuscles. Using this framework, he could argue that the incorporation of one substance into another involved neither transformation nor destruction at the corpuscular level. The aggregate corpuscles making up human bodies remained present through the whole process, merely concealed by being temporarily arranged with those of other substances to produce different sensory effects. By way of example, he cited cases where sensory evidence seemed to indicate the continued presence of aggregate corpuscles that had been incorporated into other substances, noting for instance that the purple color of juniper berries could be imparted to the flesh of pigs that fed on them.[45] Such examples, however, only represented half of the process presumed necessary. Boyle needed a visible phenomenon capable of evidencing not just the persistence of aggregate corpuscles in other substances but also the possibility that they could be recomposed just as they had been before.

Boyle found this phenomenon in a chemical experiment known as the "reduction to the pristine state." A decade beforehand, he had made prominent use of a version of that experiment involving camphor in *The Origine of Formes and Qualities* (1666). In that influential work, he had used experimental observations both to attack scholastic theories of matter and to urge the acceptance of his mechanical, corpuscular hypothesis instead. Boyle saw the camphor reduction as a decisively powerful piece of perceptible evidence for the existence of his aggregate corpuscles and for the mechanical quality of the interactions between them. If we are to grasp the significance of Boyle's use of the camphor reduction in his discussion of the resurrection, we must therefore first examine how he used it in *Formes and Qualities* to bring inconceivably small aggregate corpuscles into the realm of sensory experience.

To begin, Boyle took a quantity of camphor and mixed it with oil of vitriol (concentrated sulfuric acid) to produce an odorless, reddish solution. As he carefully explained, this solution had none of the original qualities that could be perceived in camphor—a white color, strong smell, and brittle texture. When he added water to the mixture, however, an exothermic reaction took place. Shortly thereafter the camphor began to precipitate out of the solution and back into view:

> If into this Liquor you pour a due quantity of fair Water, you will see (perhaps not without delight) that, in a trice, the Liquor will become pale, almost as at the first, and the Camphire, that lay conceal'd in the pores of the Menstruum, will immediately disclose it self, and emerge, in its own nature and pristine form of white floating and combustible Camphire, which will fill [. . .] the Air with its strong and Diffusive Odour.[46]

Describing the experiment's outcome, Boyle carefully noted that the substance that emerged from the solution possessed the exact same perceptible qualities—color, texture, flammability and aroma—as the camphor had at the start. As far as the human senses could discern, the reduction to the pristine state presented the disappearance and subsequent perfect reappearance of the substance. It is perhaps not hard to see why Boyle came to see this experiment, culminating in the instant recovery of a substance from apparent annihilation, as an analogue for the resurrection itself.

William R. Newman has shown that Boyle regarded the reduction to the pristine state as a particularly effective demonstration of the corpuscular hypothesis. For Boyle, the evident reappearance of camphor at the experiment's end made it reasonable to infer that the same substance had also been present when concealed from view through its mixture with sulfuric acid. The camphor reduction thus served as an empirical demonstration that a substance had persisted in an imperceptible state. With this demonstration, Boyle could in turn argue that all such substances were composed of aggregate corpuscles—semipermanent clusters of atoms so small as to be imperceptible when disaggregated from one another and joined to those of other substances.[47] Boyle also regarded the experiment as an empirical demonstration that those imperceptible aggregate corpuscles moved and interacted in a mechanical fashion. Consider, for instance, his reasoning about the physical causes of camphor's strong scent. For Boyle, the fact that the scent of camphor could be eliminated and subsequently restored by two odorless compounds—oil of vitriol and water—suggested that odor was not a real property of bodies. He proposed instead that it was an effect worked on the senses by changes in the size, shape, motions, and resulting texture of particles as they were placed mechanically into varying combinations. He established this point by means of a markedly anthropocentric strategy. Boyle had effected a complete transformation in the perceptible qualities of camphor, including of course its scent, through the simple mechanical act of mixing it with other substances. It seemed likely to him, therefore, that analogous mechanical processes were at play among the aggregate corpuscles themselves. In the absence of such direct evidence, Boyle worked by assuming that the phenomena open to sensory experience—in this case, his own experimental interventions—could stand in for the otherwise inconceivable processes going on at the microscopic level. He inferred the workings of an invisible mechanism from the mechanical quality of visible operations that he had himself performed.[48]

Wojcik interprets Boyle's emphasis on the intelligibility of such mechanical inferences as an indication that he did not necessarily regard them as truthful explanations of the world as God had created it.[49] It is true that Boyle understood the probabilistic knowledge resulting from experiments and observations to be provisional, and therefore apt to be modified or refuted in the future. It is also true, as we saw in *Things Above Reason*, that he believed in a God quite capable of constructing the world in a manner that did not match up with either what humans had discovered of it or were capable of understanding. Despite these admissions, Boyle nevertheless argued that natural philosophy could reveal real truths about God's creation.[50] A good example of this contradictory tendency is to be found in *Formes and Qualities* itself. Boyle opened the work by announcing his intention to frame a purely hypothetical account of how chemical transformations might be brought about. Soon, however, he changed his tone, repeatedly asserting that his micromechanical account of physical causation was in fact a true statement of how nature worked. When he got to the camphor reduction, for instance, he did not stop at identifying the corpuscular hypothesis as an explanation of how changes in odor might conceivably be brought about. Instead, he bluntly stated that what the human nose understood as changes in "Odour" were caused by changes in the "Texture" of corpuscles when they came to be aggregated and disaggregated in differing combinations. By this point, Boyle was asserting the truth of the corpuscular hypothesis, arguing that imperceptible aggregate corpuscles operated according to the same principles observable in phenomena more readily accessible to sensory experience.[51]

In *Possibility of the Resurrection*, by contrast, Boyle stuck to the more modest aim of obtaining a plausible account of how an occurrence such as the resurrection might be brought about. He sought, as we have seen, only to establish that such an event was a possibility—that it did not run contrary to reason even though an exact or thorough explanation for how it would take place could never be obtained. As Boyle saw it, the way to accomplish this end was by showing not only that processes such as the resurrection lay open to observation but also that those processes were in fact readily accessible to sensory experience. He therefore ran through several potential analogues for the resurrection, including experiments purporting to show the palingenesis of plants from their burnt ashes, before finally coming to a detailed account of the camphor reduction.[52] As in *Formes and Qualities*, Boyle once again held up the camphor reduction as a demonstration that substances were composed of infinitesimally small aggregate corpuscles and that perceptible changes in their qualities came about through mechanical alterations to the configuration of those corpuscles. The camphor's reappearance at the experiment's end was therefore to be understood as the result of using water to mechanically separate the camphor corpuscles from those of sulfuric acid and to return them to their original configuration.[53]

Equipped with experimental evidence for a similar process, Boyle could move on to establish the possibility of the resurrection itself. He proposed that God could, if he so chose, use such a chemical method to reconstitute each and every human body ahead of Judgment Day:

> *Since*, I say, these things are so, why should it be impossible, that a most intelligent Agent [. . .] should be able so to order and watch the Particles of a Humane Body, as that *partly* of those that remain in the Bones, and *partly* of

> those that copiously flie away by insensible Transpiration, and *partly* of those that are otherwise disposed of upon their resolution, a competent number may be preserved or retrieved; so that stripping them of their disguises, or extricating them from other parts of Matter, to which they may happen to be conjoined, he may reunite them betwixt themselves, and, if need be, with particles of Matter fit to be contexted with them, and thereby restore or reproduce a Body, which, being united with the former Soul, may, in a sense consonant to the expressions of Scripture, recompose the same Man, whose Soul and Body were formerly disjoined by Death.[54]

With these words, Boyle pictured the resurrection as a more complex version of the camphor reduction. He imagined God as a transcendently skilled chemist, reversing the decomposition of human bodies by extricating their aggregate corpuscles from the new substances into which they had been incorporated and then reconstituting them in their original configurations. If a human could recover camphor from its apparent destruction through a series of simple mechanical interventions, then it stood to reason that a far wiser and more powerful agent, God, could accomplish the same thing on a much grander scale.[55]

Boyle made the resurrection into a partial object of knowledge using a strategy entirely consistent with his approach to the representation of imperceptible particles. In each case, he asserted that objects accessible to sensory experience could be made to stand in for entities that would otherwise remain inconceivable. In other words, he asserted that comparisons between imperceptible entities and the objects of sensory experience were both possible and capable of yielding reliable insights. Recall that Boyle's interpretation of the physical causes of the camphor reduction depended on a presumption of ontological similarity not only between perceptible and imperceptible phenomena but also between his own experimental interventions and the processes he believed to be going on at the corpuscular level. Although Boyle sought a lower grade of certainty when it came to the resurrection, he nevertheless used the same strategy to make it into an object of knowledge. The only difference was that he extended the ontological scope of the comparison further, applying it not only to the experiment's physical causes but also to the mysterious processes involved with the resurrection itself. The entire premise of his argument was that the imperceptible physical processes supposedly revealed by the experiment could be made to somehow stand in for and yield knowledge of the supernatural, unintelligible event promised in the Bible. Boyle's outwardly modest claim to be demonstrating nothing more than the possibility of the resurrection concealed a remarkably ambitious attempt to demonstrate the theological potential of experimental natural philosophy.

An Empiricism of the Imagination

Boyle compared the mind's encounter with God to the eye's encounter with the ocean, the workings of imperceptible aggregate corpuscles to those of visible machines, and the doctrine of the resurrection of the body to a chemical experiment. He used those comparisons to make inconceivable entities conceivable, dragging them from the realm of imperceptibility into the realm of things accessible to the senses. For Boyle, the imagination therefore provided both the problem and its solution to the question of imperceptible entities. On one side, the inability to supply the mind with images of things inaccessible to the senses made not only the observation but even the contemplation of such entities impossible. On the other side, however, the imagination's comparative faculty—its capacity to substitute images derived from sensation in the place of others—provided a kind of answer. Using an assortment of images purportedly like the imperceptible entities in question, Boyle could claim to obtain knowledge of what would otherwise have remained outside his grasp.

None of this is to deny that Boyle's radical voluntarism put him at odds with contemporaries who also pursued physico-theology. It is safe to say that both Nehemiah Grew and John Ray, for instance, held slightly more "intellectualist" positions, regarding the world more as an emanation of God's reason and therefore more open to the human faculties. As Scott Mandelbrote has noted, both Ray and Grew were far more comfortable than Boyle with the idea that the study of nature could reveal the wisdom of God. Works such as Ray's *Wisdom of God Manifested in the Works of Creation* (1691) and Grew's *Cosmologia Sacra* eschew Boyle's emphasis on God's unlimited power and freedom of will in favor of an emphasis on the world's orderliness and intelligibility.[56] One result of this emphasis on God's reason, as Peter Harrison has shown, was that Grew adopted a markedly different position on the relationship between miracles and the laws of nature. He could not accept Boyle's suggestion that God was free simply to suspend the laws of nature to make miracles happen. Instead, he for the most part suggested that miraculous occurrences, such as the plagues inflicted on Egypt to encourage the pharaoh to release his Israelite captives, came about through extremely long chains of natural causes implanted in the world by God at the beginning of time. They were scheduled, so to speak, to coincide with moments when humanity had need of assistance, chastisement, or a message of unmistakably divine origin.[57]

Despite such differences, however, it is striking that Grew and Ray used the same strategies when it came to imperceptible entities, albeit binding the work of representing atoms even more directly to that of representing God's intentions. Drawing on his earlier work dealing with the saline chemistry of plants, for instance, Grew argued in his *Cosmologia Sacra* that the order of the atomic world could be safely inferred from the order so readily apparent to him in perceptible phenomena. Immediately afterward, however, he used the same argument to show that God was the agent responsible for creating that order:

> Now Regularity, which is certain; cannot depend upon Chance, which is Uncertain. For that were to make Uncertainty, the Cause of Certainty. [. . .] It is therefore evident, That as Matter and Motion; so the Cizes and figures, of the Parts of Matter, have their Original from a Divine Regulator.[58]

Let us spell out exactly what Grew was suggesting here. It was not just that the behavior of imperceptible-but-material atoms could be inferred from the behavior of material things open to the senses. Rather, it was that the same phenomena could also shine a light on the attributes and intentions of an imperceptible, immaterial, and infinite God. This was a position, moreover, that Ray anticipated in a letter that he wrote in 1685 to Tancred Robinson, where he similarly argued that the regularity observable in visible saline particles was a compelling argument for regularity among atoms. Both Grew and Ray assumed, therefore, that things accessible to the senses and imagination could stand in for imperceptible entities of uncertain ontological status. Like Boyle, they assumed that there was enough continuity between the perceptible and imperceptible parts of the world, including the human faculties and the mind of God, for the world of experience to stand in for the wisdom of God.[59]

Focusing on representational strategies rather than just explicit statements of belief is neither to deny nor diminish the importance of theology and metaphysics to natural philosophy in the later seventeenth century. Rather, it is to recognize that philosophers of the seventeenth century did not always share today's assumption that belief—rather than knowledge—motivates religious ideas. In this chapter, as a result, we have seen that the religious "content" of Boyle's thought is to be found not only in statements of what he believed but also in the practices with which he inquired into and represented things. For sure, Boyle's radical voluntarism led him to adopt positions in both theology and natural philosophy that differed in significant respects from those of his contemporaries. Despite such disagreement, Boyle's approach to imperceptible entities shows us that he often worked with ontological assumptions closer to those expressed by Ray and Grew. His comparisons bespeak continuity between the human faculties and the mind of God. They bespeak a world in which things intelligible as the products of sensory experience could somehow lead to knowledge about inconceivable entities of uncertain ontological status. Whether applied to material atoms or an immaterial God, Boyle's strategies for opening up imperceptible entities to empirical study expressed the same ontological assumption—namely, that there was enough continuity between the perceptible and imperceptible parts of the world for the world of experience to stand in for the wisdom of God.

In *The Christian Virtuoso* (1690), one of his last published works, Boyle sought to justify his use of comparisons in the face of some now unknown critique. He responded by arguing that his comparisons could serve the same purpose as analogies in philosophy. Mobilizing the logic that had informed his use of the camphor reduction to establish the possibility of the resurrection, he argued that comparisons could serve as explanations for the things they indirectly represented:

> Apposite Comparisons do not only give Light, but Strength, to the Passages they belong to, since they are not always bare Pictures and Resemblances, but a kind of Arguments; being oftentimes, if I may so call them, Analogous Instances, which do declare the Nature, or Way of Operating, of the Thing they relate to, and by that means do in a sort prove, that, as 'tis possible, so it is not improbable, that the Thing may be such as 'tis represented.[60]

Here, Boyle finally made it explicit that the imaginative substitution of one idea for another could lead to knowledge of otherwise mysterious phenomena. In almost the same breath, however, he acknowledged that many contemporaries would be tempted to dismiss his comparisons as images—"bare Pictures and Resemblances"—that could reveal little about the things for which they stood. Indeed, just a few lines earlier, Boyle had pointed out that the comparison was widely understood not as an instrument in the production of knowledge but rather as a weapon in the arsenal of affective, and therefore potentially misleading, devices used by poets and orators. Moreover, he admitted that his own representational praxis was informed by this understanding of the comparison, explaining that "my reasons for this Practise, were, not only because fit Comparisons are wont to delight most Readers, and to make the Notions, they convey, better kept in Memory." In future chapters, moreover, we will see that philosophers such as Willis's sometime student John Locke explicitly identified the pleasures of poetry and rhetoric with the workings of the imagination, regarding the comparison of ideas in the brain as the cause of bodily pleasures that might distract the mind from the pursuit of reason.[61]

This chapter's discussion of the entangled metaphysical and aesthetic questions raised by imperceptible entities will therefore mark the beginning of a broader reconsideration of the empirical project pursued by naturalists such as Boyle, Grew, Ray, and Willis. So far, we have seen that natural philosophy had far more in common with physico-theology than has widely been assumed. By examining some of the representational practices common to the two genres, we have uncovered the metaphysical assumptions bound up with their attempts to make imperceptible entities accessible to the senses. It is not yet clear, however, how naturalists could reconcile the pursuit of purportedly objective knowledge about the world with the use of imaginative strategies that were widely assumed to involve the provocation of apparently subjective affective states, such as sensory pleasure. Yet the pleasures of the senses were clearly important to the empirical project. In physico-theology, as we have seen, Ray and his contemporaries identified sensory pleasure as a crucial part of the experience that would lead to both knowledge of nature's design and knowledge of God. The claims of physico-theology depended, indeed, on an aesthetic ideal that regarded nature as an orderly and beautiful product of God's necessarily perfect designing agency. The difficulty, however, was that nature did not always look beautiful or elicit much in the way of sensory pleasure, nor was it always easy to identify as the work of a brilliant designer. Even the most convinced physico-theologians sometimes saw scenes of chaos and disfigurement in nature rather than the beauties for which they hoped. Our next task, therefore, will be to examine how these philosophers responded when nature failed to live up to the aesthetic ideal of beauty and order so important to their image of the world as a stable system perfectly designed by God.

3

In Search of Lost Designs

In the early 1690s, the naturalist and antiquarian Edward Lhwyd (ca. 1659–1709) wrote to John Ray, recounting some of the things that he had observed on a visit to the mountains of what is now called Caernarfonshire, in North Wales. Among the things he had seen, one seemed hard to explain:

> At the highest Parts of the *Glyder*, (a Mountain about the height of *Cader Idris*) there are *prodigious heaps of Stones*, many of them of the largeness of those of *Stonehenge*, but of all the irregular Shapes imaginable, and they all lie in as much Confusion as the Ruins of a Building can be supposed to do.[1]

The stones that Lhwyd described present the same impression today as they did in the seventeenth century. Perched close to the summits of what are now known as the Glyder Fach and Glyder Fawr mountains, these slablike gray stones resemble giant structural members that have somehow been left in a state of precarious disorder. Lhwyd would have found these formations easy enough to interpret—as he explained to Ray—if he had spotted them in a valley. Using Ray's arguments about water erosion, he would have seen them as rocks that had become dislodged from the nearby mountains by years of rainfall. Their presence at the top of the mountain, however, made them more difficult to understand.[2]

It is clear enough from Lhwyd's description, with its reference to Stonehenge and comparison to the "Ruins of a Building," that the stone formations atop the Glyderau range reminded him of architectural ruins. Perhaps this comparison occurred to him because of his already well-developed interest in what we now understand to be prehistoric monuments.[3] Lhwyd used the last page of his letter to consider the resemblance further:

> For that ever they should be indeed the ruines of some Edifice, I can by no means allow [. . .]; we must then allow them to be the Skeleton of the Hill exposed to open view by Rains, Snow, &c. but then how came they to lie across each others in this Confusion[?][4]

Lhwyd was sufficiently engaged by the similarity between the stones that he had seen and architectural ruins to explicitly deny that they were the vestiges of some ancient monument. But even as he rejected the idea that he had encountered the traces of a building, he took seriously the broader explanatory potential of the resemblance. He entertained the possibility that the stones could be interpreted as if they were the ruins of a natural structure. Perhaps they contained traces of an original, more perfect design that, like the design of Stonehenge, could be recovered and brought to light.

This chapter explores a similar preoccupation with the interpretation of ruins. What Lhwyd's letter to Ray serves to illustrate is that attempts to recover the design of apparently ruined objects linked a wide range of the Royal Society's interests, from architectural design to antiquarian research and natural history. Beginning with a focus on debates about architectural design and ending with a controversy in natural history, we will see that strategies for interpreting ruins were far more important to the day-to-day practices of natural history than has so far been recognized. The aim is therefore not simply to repeat the now familiar observation that there were links between architecture and natural philosophy in the seventeenth century. Matthew C. Hunter has recently shown that Robert Hooke saw the practices of architectural design as a powerful model for producing natural knowledge. Hooke believed that the effectiveness of this architectonic model for knowledge production could be attributed to the fact that it reflected the operations of the mind itself. J. A. Bennett has meanwhile demonstrated that Christopher Wren saw architectural design as a practice that depended on natural principles, making architecture an adjunct of natural philosophy. What is more, Wren believed that the forms of representation belonging to mathematics—including the representational practices of classical architecture—provided the most powerful language for communicating those principles.[5]

It has never been noted, however, that an interest in the interpretation of ruins was common to the Royal Society's architectural interests and its work on natural history. Yet ruins had an important role in each of these areas. For architects, ruins were valuable because they held clues about the true principles of architectural design—principles thought to have been lost, owing to the amnesia of the "dark ages." In his architectural writings, Wren thus mingled analyses of geometry and statics with restitutions of the designs of ancient monuments, from Solomon's Temple in Jerusalem to the Temple of Mars the Avenger in Rome. Meanwhile, as Lhwyd's letter to Ray reminds us, naturalists interested in the history of the earth sometimes made the comparison between natural history and architecture explicit. Indeed, it is well known that participants in the controversy over the age of the earth provoked by the publication in English of Thomas Burnet's *Sacred Theory of the Earth* (published in two volumes, 1684 and 1690), explicitly mobilized some of the strategies of architectural and antiquarian criticism as they attempted to infer the earth's previous forms from its present-day appearance.[6]

However, the comparison between architectural and natural-historical practices is instructive even for cases in which naturalists inquired into objects that—unlike the stones marking the traumatic history of the landscape—bore little immediate resemblance to monumental ruins. The search for ruins in nature was widespread in natural-historical inquiry, and the range of objects encountered by naturalists that might be classed as ruins was very wide indeed. Natural historians either knew, or on occasion chose to believe, that many of the objects that they studied were vestiges of bodies that had once exhibited a higher degree of perfection. In some cases, the causes of decay and imperfection were obvious. Many of the plants that Ray worked with were samples that had been dried and pressed between the pages of an herbarium. As he complained in his letters to Hans Sloane and others, such samples were often rather imperfect guides to the live plants that they stood in for. On other occasions, the vicissitudes of time and travel could only be guessed at. Robert Hooke, for example, believed snowflakes to be the ruins of crystals that, before their violent journey through the air, had been far more beautiful. So, the task confronting naturalists was not always that of uncovering design in things as they encountered them. Instead they often had to interpret natural bodies as if they were ruins that, like those of ancient monuments, exhibited clues about the existence of former, more perfect designs.

In the first part of this book we saw that, for Ray and his fellow naturalists, the possibility of gaining natural-historical knowledge depended on the assertion that God had designed nature and that God had done so in a manner at least somewhat commensurate with human intelligence. The physico-theological assertion of God's role in creation had aesthetic consequences—that is, consequences for how nature was supposed to look and how it should be experienced. If God had designed nature, it followed that natural things should be both beautiful and perfectly designed, with form exactly matching function. Thinking about the Royal Society's interest in ruins makes it possible to see that these aesthetic claims were of deep importance to the material and intellectual practices of natural history. The dilemma posed by ruins, whether artificial or natural, was partly an aesthetic one. There was a tension between their status as objects that were assumed to have been designed and the fact that it was hard to identify in them the qualities—beauty and utility—taken to be the signatures of good design. As a result, the practices used by architects and naturalists in their efforts to uncover the design of ruined or transformed objects are particularly illuminating. This chapter therefore focuses on several puzzles about the interpretation of ruined, lost, and transformed objects, beginning with the seventeenth-century debate about the meaning of Stonehenge and moving on to an epistolary dispute between Martin Lister and Nehemiah Grew about the structure and uses of the roots of plants. Focusing on these debates reveals that natural history was an aesthetic science, its practices of inquiry and representation shaped by the assumption that the products of design were both beautiful and perfectly suited to their purposes.

Inigo Jones and the Ruins of Stonehenge

In 1655, the architect John Webb (1611–1672) published a new interpretation of the ruins of Stonehenge. Entitled *The most notable Antiquity of Great Britain vulgarly called Stone-Heng*, the book was derived from notes left behind by Webb's recently deceased mentor Inigo Jones (1573–1652).[7] Jones was the first English architect to work in the classical style, and today he is remembered mostly for buildings such as the Banqueting House in Westminster and the Queen's House at Greenwich. In the century after *The Most Notable Antiquity* came out, however, Jones was also known for what by modern standards seems a remarkably improbable claim. Stonehenge, he argued, was a Roman monument, built according to the principles of classical architecture. While Jones's claim never commanded wide acceptance, contemporaries nevertheless took it seriously. Indeed, it prompted lengthy responses from two important members of the early Royal Society. In 1663 Walter Charleton, whom we encountered in chapter 1 as the author of the first explicitly physico-theological treatise, assailed Jones and Webb in 1663 with a book entitled *Chorea Gigantum; or, The most Famous Antiquity of* GREAT-BRITAN [. . .] *Restored to the* DANES. Offering his restoration of the monument's true meaning as a symbolic counterpart to Charles II's recent restoration to the English throne, Charleton maintained that the Danes had built Stonehenge for electing and enthroning kings during the Viking occupation of Britain.[8] Meanwhile, the antiquary John Aubrey included careful responses to Jones's interpretation in the manuscript notes for his unfinished and unpublished "Monumenta Britannica." His exhaustive fieldwork and antiquarian researches led him to believe that Stonehenge was in fact the work of the British druids described in Roman accounts of Celtic customs and culture.[9]

Although none of these interpretations have stood the test of time, scholars have nonetheless detected in them progress toward the empiricism of modern archaeology. The grounds for this assessment largely concern the differing extents to which the participants in the debate mobilized evidence derived from archaeological objects themselves—as opposed to the textual evidence that had sustained the antiquarian scholarship of the sixteenth century. Jones's *Most Notable Antiquity* is regarded as the least scientific of the lot. His representations of Stonehenge seem remote from much experience with the monument itself, while his arguments are grounded to a large extent in the authority of ancient texts, in particular the Roman architect Vitruvius's (first century BCE) *De Architectura*. By contrast, Aubrey's interpretative strategy places far more emphasis on reckoning with the material remains of antiquity, giving it a correspondingly closer resemblance to modern archaeological methods. On this basis, a clear case can be made for the Royal Society's involvement in the emergence of what became modern archaeology. As Michael Hunter has shown, it was not that Aubrey and others revolutionized antiquarian scholarship by inventing new, more empirical, forms of inquiry to replace the textual erudition that characterized Jones's work. Rather, the society's rejection of textual authority in favor of the study of real phenomena encouraged and legitimated their focus on the parts of antiquarian scholarship that already dealt with material things, such as the interpretation of ancient coins, medals, and inscriptions.[10] In this light, Jones's interpretation of Stonehenge comes across as an example of precisely the kind of textual erudition that the Royal Society hoped to replace with knowledge drawn from experience.

However, the Royal Society's empiricism should not, I think, be characterized simply as an attempt to gain knowledge through the direct or unmediated experience of nature. Ray and his fellow naturalists worked under the assumption that natural things had been designed and that their inquiries were going to reveal how those things manifested the purposes of a designing agent. It is with this point in mind that Jones's interpretation of Stonehenge proves an illuminating counterpart to the practices of natural history. Like many works of natural history, it was conceived as a demonstration that the intentions of a designer could be recovered from a material thing. Indeed, Jones and Webb positioned *The Most Notable Antiquity* as an explicit rejection of the scholarly strategies that have since been attributed to them by many historians of archaeology. Jones argued instead that it was only by mobilizing the representational strategies of architectural design that the meaning and uses of Stonehenge could be recovered. He thus explained toward the end of the book that only architecture furnished the resources necessary to interpret the monument: "whosoever goes about to enforce other reasons, do as I conceive but beat the air, neither can they reduce this *Antiquity* to any probable Originall."[11] It was this same claim that motivated Jones in his much-criticized reliance on Vitruvius's *De Architectura* when he came to explain why the monument had been built. Having concluded that Stonehenge was the work of Roman architects, it made sense to find out its meaning by comparing it to the examples of architectural design that Vitruvius had described.[12]

Yet it was easy to object to this comparison. In his *Chorea Gigantum*, Charleton made the obvious point that Stonehenge did not look much like a Roman monument, remarking that its stones were far too "rude, rough, craggy, and difform among themselves" to be the products of architectural design.[13] Jones and Webb, however, had already preempted this criticism by making the reasonable argument that the monument must have suffered a great deal of violence in its long history—enough to change its outward appearance considerably. Jones complained in *The Most Notable Antiquity* that some of the smaller uprights had even been moved around in the short time between the two surveys of the site that he had carried out himself, inviting readers to extrapolate from this observation a longer history of damage and transformation.[14] Consequently, he repeatedly insisted that his task was not to represent the ruined version of Stonehenge that visitors could see with their own eyes. Instead, he was going to reveal the monument's original design—the design that would reveal the true intentions of its architect:

> Which, that it may be more clearly demonstrated, (being by me, with no little pains, and charge measured, and the foundations thereof diligently searched) I have reduced into *Design*, not onely as the ruine thereof now appears, but (as in my judgment) it was in its pristine perfection.[15]

With his telling use of the term *design* to describe a technical drawing, a usage only recently imported into English from Italian texts on art and architecture, Jones signaled his intention to represent Stonehenge with a series of technical drawings drawn from the canons of classical architecture. In other words, he asserted that the only way to recover the monument's "pristine perfection" was by turning to the intellectual and technical resources of architectural design.[16]

Jones's ambition went far beyond merely recovering the original form of Stonehenge. With his architectural drawings, he set out to get into the mind of the monument's original architect, reproducing each step in the process by which he had worked. He thus opened with two plans, one depicting the whole site and the other depicting the stones in their original situations. Of all the drawings that Jones offered up, the latter of the two plans was by far the most important. This was the plan on which architects were supposed to inscribe their initial conception for the whole building, in the form of a symmetrical outline of the positions of its walls, entries, exits, and passageways. It was on this conception, moreover, that the rest of the design would be based.[17] After this fashion, Jones realized the layout of Stonehenge as a series of regular geometrical forms (figure 2), comprising two outer circles and two inner hexagons, linking those forms together with four equilateral triangles.

From here, Jones proceeded with the next two drawings in the sequence that classical architects were supposed to follow. The first of these was the

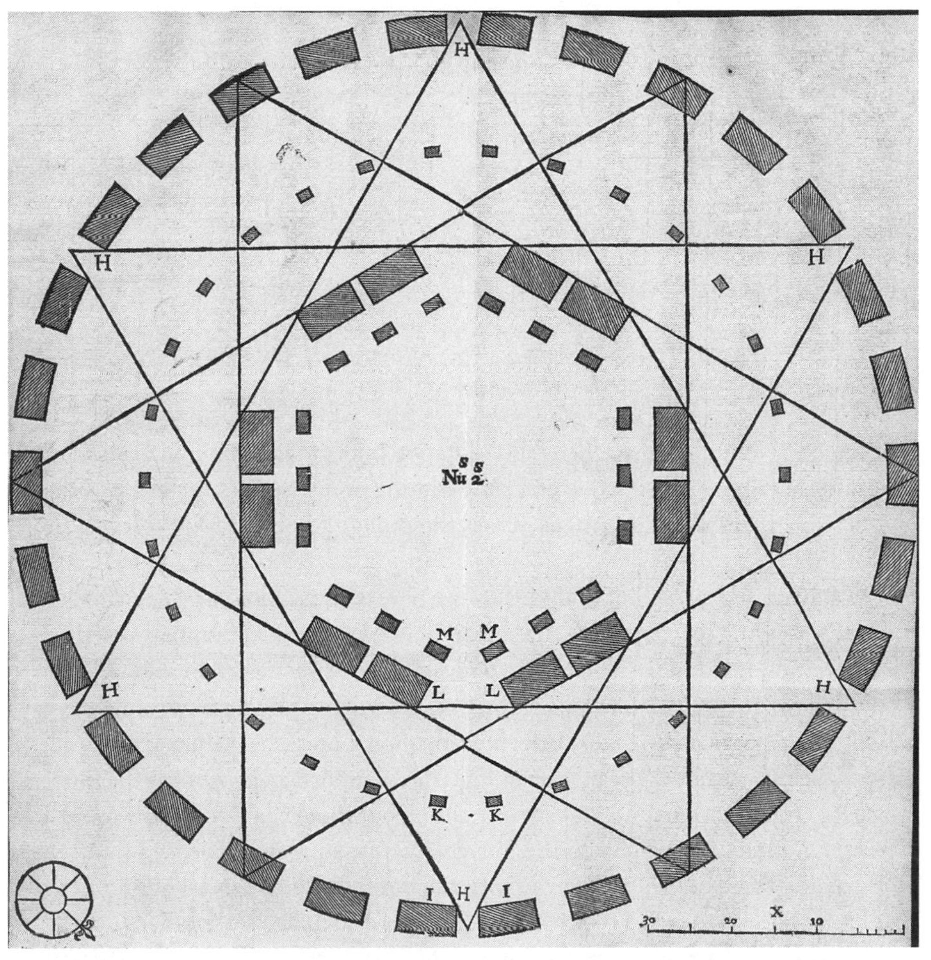

FIGURE 2. Inigo Jones's plan of Stonehenge, inscribed with four equilateral triangles forming a hexagon. Plate 2 from Jones and Webb, *The Most Notable Antiquity*. By kind permission of the Syndics of Cambridge University Library.

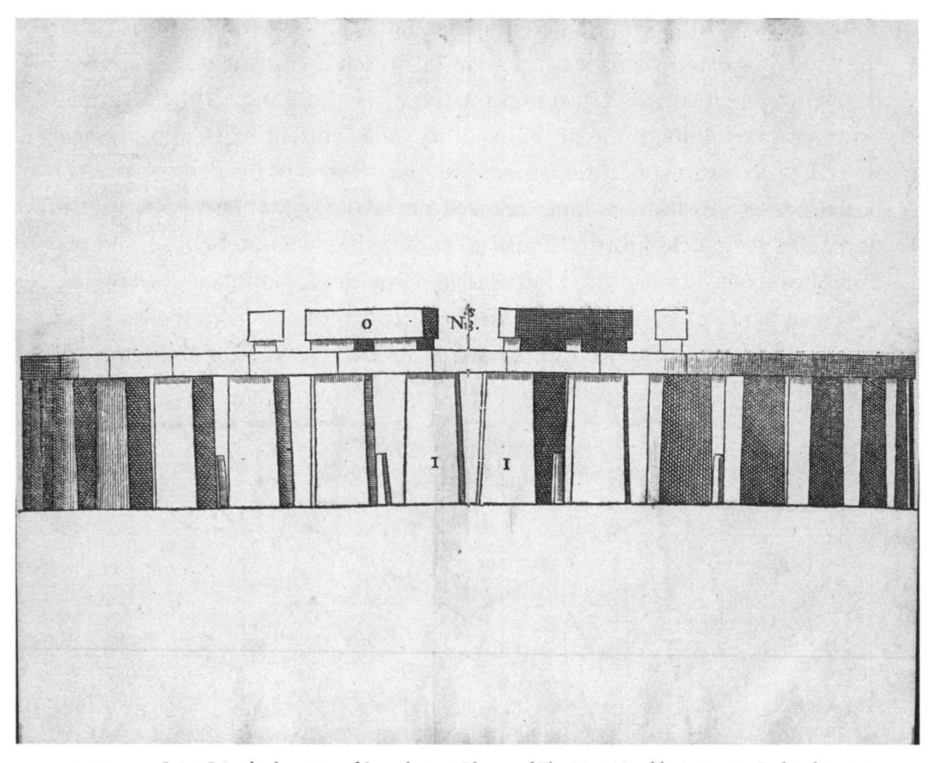

FIGURE 3. Inigo Jones's elevation of Stonehenge. Plate 3 of *The Most Notable Antiquity*. By kind permission of the Syndics of Cambridge University Library.

elevation (figure 3), a flattened representation of the building's frontal surface, realized without perspective and only hinting at depth. The second was the section (figure 4), a slice made across the building horizontally to reveal its innards.

Having prepared the ground with the necessary architectural drawings—corresponding to each of the three spatial dimensions—Jones finally used an illustration realized in perspective to reveal the Roman design for Stonehenge (figure 5). The monument was composed of four concentric rings of massive columns, the two closest to the center forming a portico-like hexagonal cell, and the two closest to the edge forming a circular portico environing the inner space. The outermost ring, as Jones pointed out more than once, was topped with an architrave joined to the uprights with tenon and mortise joints—a sure sign that a skilled Roman architect had directed the construction.[18]

Although Jones subsequently added two drawings of Stonehenge as he thought it had actually appeared, it was this sequence of architectural designs that furnished the most important evidence. He felt able to assert that the monument was a piece of Roman architecture on the basis of the ground

FIGURE 4. Inigo Jones's section of Stonehenge. Plate 4 of *The Most Notable Antiquity*. By kind permission of the Syndics of Cambridge University Library.

plan alone. The geometrical forms inscribed in this plan—a hexagonal cell environed by a circular portico—could never have been the work of the ancient Britons, a people whose barbarism was plain from Roman accounts like those of Julius Caesar and Tacitus. Among Britain's ancient inhabitants and occupiers, only the civilized Romans had possessed enough skill to plan out a building with so "much *Art*, *Order* and *Proportion*."[19] Moreover, the operation of reducing Stonehenge to plan, section and elevation enabled Jones to go far beyond this initial attribution, leading him to form strangely precise notions about the monument's original purpose. The use of formal architectural drawings transformed Stonehenge from a ruin into an architectural object, facilitating far easier comparison to other such objects. The priority accorded to the plan in the design process led Jones to devote much of his discussion to the identification of counterpart buildings that exhibited a similar geometrical layout. He did so by turning to Vitruvius's *De Architectura*, identifying two such structures—the *monopteros* and the *peripteros*, both of which were temples environed by a circular portico comparable to the one that could be found in the plan for Stonehenge. Having concluded that the monument

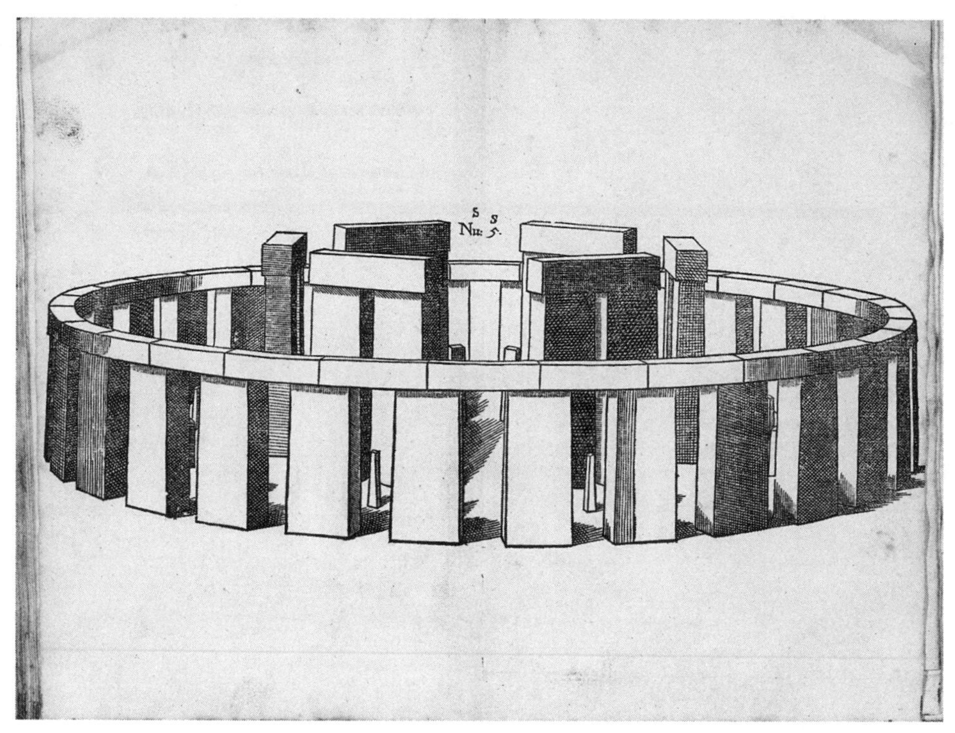

FIGURE 5. If Stonehenge had been built by a Roman architect? Inigo Jones's perspectival realization of Stonehenge. Plate 5 of *The Most Notable Antiquity*. By kind permission of the Syndics of Cambridge University Library.

must have been a kind of temple, Jones only needed a few more deductions derived from the proportions of the rather thick columns to show that Stonehenge had been designed as a temple dedicated to Caelus, the Roman god of the sky.[20]

There is obvious circularity in Jones's argument. Indeed, his attribution relied in large part on asserting that drawings depicting the "original" design of Stonehenge could supply a uniquely powerful insight into the its meaning. For Jones, this was because the plan, section, and elevation reproduced the very process by which Stonehenge had been designed, giving him an almost unmediated form of access to the intentions of its architect. Scanning the monument in three axes for the placement of walls, columns, openings, and passageways, these drawings privileged the geometrical forms through which architects working in the classical manner encoded their intentions. At the same time, they left out a huge quantity of information—for instance, concerning color, decoration, and materials—seen by classical architects as less important to understanding the design. The insights to be gained from Jones's drawings therefore depended on the assumption that the architects of Stonehenge had gone about encoding their intentions into the building in the same manner. We can also identify this kind of reasoning in Jones's claim that the present-day ruins of Stonehenge must once have been formed into symmetrical circles and hexagons. Given that those forms were not so readily apparent in the visible ruins, the argument depends on the assumption that the monument's Roman designers could never have envisaged anything other than a plan based on regular geometrical forms. But as Aubrey and Charleton were quick to point out, the validity of this representational strategy, along with the comparisons and deductions facilitated by it, depended on Jones's opening assertion that the architect had been a Roman working in the classical manner. The possibility remained that Stonehenge might have been the work of altogether different builders, whether Aubrey's British druids or Charleton's Vikings, to whom classical architecture was unknown. In such a case, there was no great likelihood that a series of architectural drawings prepared according to precepts that would have been understood by Vitruvius could yield insights of any value into the intentions of the monument's designers.

Robert Hooke and the Ruins of Snowflakes

Inigo Jones's interpretation of Stonehenge depended on assumptions about the behavior and intentions of its designer. He recognized that the present-day form of the monument manifested imperfections when judged by the standards of classical architecture. Nevertheless, he believed that Stonehenge was a piece of classical architecture—that it was the work of a designer who could under no circumstances have intended those imperfections. The only way to explain the present-day form of Stonehenge was to try to demonstrate that its designer had originally intended something far better. Jones therefore used the plan, section, and elevation both to recover and to communicate this original, more perfect design. Such a strategy might at first seem to reveal little about the practices of natural history. Jones's lack of interest in reckoning with the ruins of Stonehenge is difficult, after all, to square with the empiricism generally thought to be so characteristic of the work of naturalists like Grew, Hooke, and Ray. In Jones's deep conviction that he knew the identity of the designer responsible for Stonehenge there is, however, a crucial point of similarity. Like Jones, those naturalists believed that they knew who the designer was. What is more, the identity of this designer made the need to account for apparent imperfections even more pressing. For there was no way that God, the omniscient creator of all things, could be held responsible for seemingly imperfect or incomplete designs.

As we began to see in the introduction to this book, the aesthetic dilemma arising from this insistence on God's role in designing nature has a central place in one of the most famous texts associated with the early Royal Society—Robert Hooke's book of observations made with magnifying glasses, *Micrographia; or, Some Physiological Descriptions of Minute Bodies* (1665). In the preface, Hooke made it plain that God was not to be blamed for the appearance of imperfections in nature. Instead, following a line of reasoning that also interested Joseph Glanvill and John Locke, he suggested that the possession of sensory organs capable of accurately perceiving nature's design was one of the perfections that humanity had lost because of the first sin. It was entirely the fault of nature's observers if, with senses corrupted by an innately sinful disposition, they were incapable of seeing the order and wisdom of God's creation.[21] Hooke's answer to the fallibility of the senses was to propose that philosophers could "recover some degree of those former perfections" enjoyed by Adam by turning to new strategies and instruments for inquiring into and representing natural phenomena. In this sense, Hooke offered up the microscope as an instrument of simultaneous sensory and moral regeneration that could, if handled with care and patience, give humanity the ability to once more apprehend the beauty and wisdom of creation.[22] The *Micrographia*'s famous illustrations drive this point home, revealing the seemingly contemptible bodies of such creatures as fleas and lice to be beautiful and perfectly designed. Meanwhile, the most refined products of the human faculties, whether the point of the needle or the shimmering texture of watered silk, are revealed to be full of imperfection.[23]

Hooke resolved the tension between the appearance of imperfection in nature and the belief in God's designing agency by appealing to the new strategies of inquiry and representation promoted by the Royal Society. Those practices, he claimed, could engender a new and better experience of nature, accompanied by a higher degree of sensory pleasure and deeper insights into God's wisdom than had been possible since Adam's fall from grace.[24] It is therefore clear that Hooke himself identified a close connection between the work of natural philosophy and the bundle of affective and moral gains to be had from the experience of purposiveness in natural bodies. As we saw in chapter 1, however, the most prevalent scholarly interpretations still regard such physico-theological arguments as distinct from the actual work of natural philosophy. On this view, Hooke's insistence that the work of natural philosophy could lead to deeper knowledge of God was simply part of a broader apologetic strategy intended to deflect the accusations of irreligion made at the time by critics of the Royal Society.[25] The evidence of Hooke's interest in what he thought to be the ruins of natural phenomena points, however, in the opposite direction. One such case reveals how much Hooke's strategies of inquiry and representation could be motivated by assumptions about the appearance of good design. Those strategies, moreover, bear a striking resemblance to the ones used by Inigo Jones in the interpretation of Stonehenge.

The ruins in question are to be found in a section of the *Micrographia* dealing with the role of crystallization in the formation of inanimate bodies. This section in turn formed part of an audacious attempt to reduce seemingly complex phenomena such as the coagulation and separation of substances to a series of relatively simple mechanical principles, based on the interactions between variously shaped invisible particles of matter. In this context, crystallization interested Hooke because he regarded it as one of the most basic and fundamental processes by which natural bodies were formed and generated. Hooke and the majority of his contemporaries no longer shared the earlier belief, central to Paracelsian medicine and much Renaissance alchemy, that saline crystals were themselves fundamental particles of matter. Indeed, Hooke thought that crystalline bodies were at least two steps removed from the invisible particles to which he assigned ultimate responsibility for the conformation of natural bodies. As he explained in the *Micrographia*, he regarded crystalline bodies as the products of different combinations of globular portions of matter that themselves resulted from the mechanical interactions of variously shaped invisible particles.[26] In spite of this crucial difference, naturalists nevertheless sought to integrate the saline chemistry into the explanatory frameworks provided by their new theories of matter. Hooke's *Micrographia* was therefore just one among several works, including Nehemiah Grew's lectures on plant anatomy and Thomas Willis's aforementioned *Cerebri Anatome*, that sought to explain the generation of natural bodies with reference to the interactions of saline bodies and the processes of crystallization.[27]

Hooke evidently thought that crystallization was one of the main processes used by God to bring natural things into being. It is perhaps reasonable to suppose, therefore, that he expected the microscope to show that crystals were particularly beautiful examples of divine workmanship. In the *Micrographia*'s discussions of mineral and salt bodies, this supposition holds true. When Hooke placed those bodies under the microscope, he observed a host of regular, geometrical crystals that were for the most part invisible to the naked eye.[28] When, however, he turned to some crystalline bodies that resulted from freezing—such as snowflakes and frost formed on the surface of glass—he had the opposite impression. Forms that seemed beautiful to the naked eye seemed far less attractive when viewed with the microscope:

> The parts of those curious branchings, or *vertices*, that usually in cold weather tarnish the surface of Glass, appear through the *Microscope* very rude and unshapen, as do most other kinds of frozen *Figures*, which to the naked eye seem exceeding neat and curious, such as the Figures of *Snow*, frozen *Urine*, *Hail*, several *Figures* frozen in common Water.[29]

In the detailed discussion of snowflakes that follows, Hooke repeated the observation. In this place, however, he confronted its potential to disrupt his aesthetic criterion for distinguishing between natural and artificial things. The ugliness apparent in snowflakes when viewed with the microscope gave them an unsettling kinship with the products of human art:

> Observing some of these figur'd flakes with a *Microscope*, I found them not to appear so curious and exactly figur'd as one would have imagin'd, but like Artificial Figures, the bigger they were magnify'd, the more irregularities appear'd in them.[30]

From here Hooke moved quickly to foreclose the disturbing possibility of this revelation—that God's designing agency was less perfect than it ought to be. There was no question of blaming the imperfection that he had witnessed on some defect in God's workmanship.

Instead, Hooke identified a more mundane, if equally speculative, solution to the problem. The irregularity observable in snowflakes was not to be ascribed to any fault on the part of their designer but instead to the "thawing and breaking" they had endured on their long descent from the clouds to the earth. Following this line of reasoning, Hooke proposed that if snowflakes could be inspected under magnification at the point of their generation in the clouds, they would appear just as beautiful as any other product of nature:

> I am very apt to think, that could we have a sight of one of them through a *Microscope* as they are generated in the Clouds before their Figures are vitiated by external accidents, they would exhibit abundance of curiosity and neatness there also, though never so much magnify'd.[31]

Here the similarities between Jones's interpretation of Stonehenge and Hooke's interpretation of snowflakes become pronounced. By identifying the snowflake as a ruin and placing its beauty at a time in the past, Hooke could preserve his assumptions about how and with what degree of perfection its designer had worked. What is more, he set out to recover this lost design. As we saw in the previous section, Jones accomplished this end by using architectural forms of representation to subsume Stonehenge under the paradigm of classical architectural design. Hooke worked in a similar fashion, more or less ignoring the visible imperfections that he had seen in favor of emphasizing the geometrical design that they seemed to obscure. As in the case of Stonehenge, this strategy made it far easier to compare snowflakes to a class of objects that Hooke took to be better preserved, more perfect examples of the same design paradigm.

Let us consider Hooke's representations of snowflakes in more detail, both those that he added to the Society's Register Book in December 1662 and those he subsequently included in modified form in the *Micrographia* itself. Accompanying the description in the Register Book is a striking ink wash drawing that depicts snowflakes on a "piece of black Cloth, or a black Hatt" (figure 6).

The strange thing about this image is that it manifests little interest in precisely visualizing snowflakes in their external aspect. Were it not for the ingenuity of the execution (areas of the ink wash have been left clear to make the surface of the paper stand in for the whiteness of the snowflakes), these blotchy snowflakes might even seem to be the work of a child. As Matthew Hunter has shown, however, a closer inspection of the drawing in the Register Book reveals a subtler strategy at work. Before applying any ink to the page, Hooke incised a series of equally sized circles into the page using a compass and razor. Next, he divided each circle into six equal parts using a straight edge and razor, as if to construct hexagons within them. It was upon

FIGURE 6. Robert Hooke's drawing of snowflakes, included with "Figures Observ'd in Snow by Mr. Hook," Royal Society Register Book, vol. 2, p. 62. Royal Society Centre for the History of Science. © The Royal Society.

FIGURE 7. Detail of Hooke's drawing from figure 6, showing the snowflake at right center.

these incised lines, placed at sixty-degree angles around a compass point, that Hooke painted the snowflakes, using the lines constructed in the circles as guides for the branches (figure 7).[32]

It must be noted here not only that Hooke had identified a good deal of irregularity in the outward appearance of snowflakes but also that in his description he parenthetically acknowledged many of them to have melted to the point that no figuration of any sort whatsoever could be identified.[33] In spite of this acknowledgement, it is clear that he chose a representational strategy designed to represent snowflakes as regularly formed bodies. Since Hooke used incised geometrical constructions to guide the painting of the snowflakes, the drawing was bound to show that their rough exterior surfaces obscured an underlying order of regular, geometrical forms.

The Register Book drawing enables us to see that the implications of Hooke's aesthetic argument went far beyond the apologetic function so widely attributed to it by his interpreters today. In fact, his assumptions about the perfection of divine workmanship informed the visual and material strategies through which he sought to uncover the principles of the formation of bodies. This point is reinforced if we place the drawing in the broader context of the *Micrographia*'s interpretation of the formation of crystalline bodies. By emphasizing their underlying regularity over their surface roughness, Hooke made it possible to form comparisons between snowflakes and other regularly formed crystalline bodies. The most important of these were the crystalline forms visible in the surface of frozen urine, a description of which immediately precedes the account of snowflakes. As Hooke explained there, he froze the urine himself under controlled circumstances, taking care to ensure that the formation of the crystals was not disturbed by gusts of wind and the like.[34] While urine crystals exhibited the same basic conformation as snowflakes—six main branches disposed on a hexagonal plan—there was a crucial difference under microscopic observation. The branches of the crystals in frozen urine, unlike the rough-hewn branches of snowflakes, seemed remarkably beautiful to Hooke. Indeed, he went so far as to write that the "exactness and curiosity of the figuration of these branches, was in every particular so transcendent, that I judge it almost impossible for humane art to imitate."[35] In other words, the crystals visible in frozen urine exhibited those qualities of beauty and perfection that Hooke expected from all products of divine design. These were the qualities, moreover, that he expected to be able to find in snowflakes if he could only examine them under like circumstances, free of any kind of damage at the time and place of their generation.

When Hooke came to make his case for the former perfection of snowflakes, he took the regularity and beauty observable in urine crystals, along with mineral and saline crystals, as the pattern for their original design. Proposing that the same generative process responsible for those crystals was also likely to be the cause of snowflakes, he submitted that it was rational to believe that snowflakes must once have exhibited the same degree of regularity and beauty. Although there is a commonsense quality to the inference that Hooke was drawing here, his argument nonetheless exhibits a circularity that echoes Inigo Jones's approach to the ruins of Stonehenge. Hooke's argument that snowflakes were the outcome of the same generative process responsible for all crystalline bodies was largely a function, as we have seen, of his representational strategy—a strategy that ignored surface irregularity as well as examples of snowflakes that were wholly irregular. And when Hooke sought to explain these irregularities, he ran the argument in the opposite direction, using the claim that snowflakes resulted from the same generative principles as other crystalline bodies to suggest that their original appearance had been far more beautiful. With their irregular appearance, snowflakes and other ice crystals presented Hooke with evidence that challenged his assumptions about the beauty and perfection of God's designing agency. The *only* frozen crystals that conformed to his expectations about the appearance of natural things under the microscope were those that he had made himself under controlled conditions. Yet it was those crystals that Hooke held up as the pattern for the development of all crystalline bodies caused by freezing. His response to the lack of beauty in snowflakes was not, then, to faithfully record what he saw. Instead he mobilized strategies of inquiry and representation behind an almost entirely imagined history of beauty and perfection ruined by external forces.

Design and the Work of Description

The case of Hooke's interest in snowflakes suggests that the Royal Society's project of representing nature was not as straightforwardly empirical as has so often been claimed. We have seen that the aesthetic argument of the *Micrographia* cannot be dismissed as a defensive addition to an otherwise scientific description of experiments and natural things. Instead, that argument informed Hooke's natural philosophy, leading him to pursue strategies intended not only to recover the lost beauty of snowflakes but also to reveal their part in a broader paradigm of divine purposiveness. It is equally clear, however, that the *Micrographia* was an unusual book. No other work of natural history or anatomy published in the Royal Society's ambit was quite so explicit in its physico-theology, quite so certain of the geometrical order in the composition of natural bodies, nor backed up by so compelling a scheme of visual rhetoric. That said, it is nevertheless possible to generalize from the *Micrographia*'s example. Here three kinds of sources prove particularly useful: the descriptions filling contemporaneous works of natural history and anatomy, the discussions about how best to represent nature in the correspondence of naturalists such as John Ray, and finally programmatic statements about the value and purposes of representation in the prefaces to works such as Nehemiah Grew's *Musæum Regalis Societatis* (1681). Together, they reveal that Hooke's approach to recovering design through the representation of nature was by no means unique. His contemporaries—naturalists such as Grew and Ray—agreed with him that the disclosure of design in nature was one of their main tasks. Far from setting out to merely record and reproduce sensory experience, they used descriptive practices with the intention of recovering the purposiveness that they believed had been designed into every natural body.

Yet there can be no denying that several of the Royal Society's members, including Hooke himself, called for an approach to representation that is much easier to square with their avowed empiricism. Indeed, Hooke used the preface of his *Micrographia* to argue that natural history required a form of representation wholly at odds with the ones that he had used in the production of the book. This was to be the mere reproduction or transcription of sensory experience, a task that he disingenuously claimed would require little in the way of intelligence or judgment:

> There is not so much requir'd towards it, any strength of *Imagination*, or exactness of *Method*, or depth of *Contemplation* (though the addition of these, where they can be had, must needs produce a much more perfect composure) as a *sincere Hand*, and a *faithful Eye*, to examine, and to record, the things themselves as they appear.[36]

These words carry an echo of the descriptive approach to natural history proposed by Francis Bacon around half a century earlier. As is well known, Bacon had argued that long and detailed descriptions of natural phenomena were required for the task of replacing the Scholastic philosophy of nature with a more truthful account of physical causation. For Bacon, such descriptions would do more than provide the new empirical foundations on which the new natural philosophy was to be built. By calling attention to aspects of natural phenomena other than those that had traditionally been deemed important, they would also free the mind from the inherited prejudices of Scholasticism. For this reason, both Peter H. Miller and Svetlana Alpers argue that Bacon's lengthy descriptions entailed a philosophically productive "suspension of judgment." Rather than choosing what to represent on the basis of what seemed important according to received wisdom, the naturalist would instead describe as many details as possible in case any of them subsequently turned out to be of importance.[37] In her influential book *The Art of Describing*, Alpers identifies Hooke's insistence that naturalists should simply transcribe sensory experience as an endorsement of this Baconian mode of description, along with its implicit suspicion of the premature exercise of judgment.[38]

It is difficult, however, to find evidence of such an indiscriminate descriptive mode in action. Instead, we find that naturalists of the later seventeenth century generally employed one of two descriptive modes, which I have termed *classificatory* and *anatomical*. The first I call classificatory because its purpose was to enable readers to distinguish one kind of thing from another. We mainly find descriptions of this sort, therefore, in works of natural history that divide the natural world into species, such as Ray's histories of birds, fish, and plants. Even though those descriptions were sometimes long and detailed, they can hardly be thought of as indiscriminately detailed representations of sensory experience. As Ray explained himself, the task of representing interspecific difference was one that exacted much judgment from the naturalist, who needed to strike a delicate balance between including enough details to make differences clear and including so many as to leave the reader overwhelmed and confused. In a letter to Hans Sloane written in

1698, for example, Ray praised the descriptions in Paul Hermann's *Paradisus Batavus* (1698), an account of the plants in the botanical garden at Leiden, for their concision. He remarked with approval that they were "withal concise, and not encumbered with superfluous and unnecessary stuff, which obscures rather than illustrates."[39] At other times, Ray gently upbraided his collaborators when they described things in too much detail—for example, adding an annotation to his *Ornithology* criticizing his deceased patron and collaborator Francis Willughby for including inconsequential details in the description of a woodpecker: "I think it is not needful so scrupulously to describe every particular spot in each feather: for that nature takes a latitude [. . .] in these lesser things, not observing always the same number, figure, and situation of spots."[40] The task of inscribing natural kinds in God's static order—the great chain of being—was to be accomplished not through the absence of judgment, but rather through its exercise.

I call the second kind of description anatomical because it reflects an approach characteristic of contemporary anatomical works—that of drawing close connections between the forms of animal bodies and their uses. Perhaps the most extensive contemporary reflection on such descriptions is to be found in the preface to Nehemiah Grew's *Musæum Regalis Societatis* (1681), his catalog and description of the Royal Society's collection of "natural and artificial rarities." Grew responded to the potential objection that his descriptions were far too long with the argument that such descriptions could be useful even when the object in question was well known. Citing an effect of defamiliarization that has occasionally figured in modern discussions of description in literary aesthetics, he wrote that "in such Descriptions, many Particulars relating to the Nature and Use of Things, will occur to the Authors mind, which otherwise he would never have thought of."[41] With their emphasis on the role of description in distancing the mind from what seems familiar, these words once again seem to follow the outlines of the Baconian descriptive project. However, Grew's insistence that this process would result in the discovery of "the Nature and Use of Things" marks a significant departure from that project. In the *Novum Organum* (1620), Bacon had argued that there was no place for arguments about final causes in the reformed natural philosophy, characterizing them as hopelessly anthropocentric.[42] By contrast, Grew made it clear that the whole point of describing things was the discovery of purposiveness in nature. In fact, he followed up this initial remark with a list of some of the examples of purposiveness to be found by attending to the small details recorded in long descriptions. It included a digression on the connection between the shapes of the ears of dogs and their hunting habits—pointed ears for those reliant on their hearing, and downward-hanging for those more dependent on their sense of smell.[43] With its self-conscious appeal to a gentlemanly audience and the simplicity of its argument, this example would not be out of place in a work of popularizing physico-theology like Ray's *Wisdom of God*.

When such examples appear in physico-theology, scholars are quick to point out their teleological character. Yet it is not often enough observed that arguments of the same kind had a central role in virtually every work of anatomy and physiology of the later seventeenth century.[44] This oversight can be explained by the fact that the arguments connecting form with function so widespread in these works are not much of a shock to modern sensibilities. After all, such assertions are still used today as commonsense descriptions of the functions of parts of the body—the purpose of the kidneys is said to be the removal of waste products from the blood, and the function of the stomach is said to be digestion of food. Ancient and early modern naturalists were, by contrast, far more open about the role of teleology in the work of anatomy.[45] During the first half of the seventeenth century, anatomists often made use of the Aristotelian form of teleological reasoning against which the mechanical philosophers so strongly objected. Peter Distelzweig and Don Bates have both shown, for instance, that William Harvey (1578–1657) did not admit the mechanisms of the circulatory system as its causes even though he made frequent use of mechanical models to describe them. Instead, he preferred a broadly Aristotelian explanation based on a "faculty" or "force" for motion located in the heart.[46] In the next few decades, however, anatomists increasingly sought to frame their arguments about the uses of the parts in the mechanical terms with which we are already familiar. As Domenico Bertoloni Meli has explained, leading anatomists of the later seventeenth century, such as Marcello Malpighi (1628–1694), sought to show that the mechanisms found in plants and animals could explain how they came to serve given purposes.[47] Like several of their contemporaries in England, Grew and Willis facilitated this explanatory model by organizing published anatomical works so that the descriptions of those mechanisms came prior to separate sections dealing with their uses. In so doing, they mimicked a format that had been in widespread use in books on scientific and navigational instruments since the sixteenth century.[48]

Arranging his anatomies into separate sections dealing with description and then use did not mean, however, that Grew sharply divided the work of description from that of making judgments about purposiveness. On the contrary, his descriptions continually anticipate and even on occasion slip into the discussion of uses. Consider, for example, the description of poppy heads to be found in the part of Grew's *Anatomy of Plants* (1682) dealing with the anatomy of fruit. Grew continually hinted at the uses that he would later discuss in more detail, ending the description by remarking that the poppy head was "A *Fabrick* designed for several purposes, as shall hearafter be said." As he described the individual parts in question, moreover, he continually likened them to human-made objects that would have been familiar to his readers. So, he described its structure as "a little *Dove Coat* [Cote]; divided by Eight or Ten *Partitions*, into so many *Stalls*," noting that "as it dries, it gradually opens at the *Top*, into several *Windows*, one for every *Stall*."[49] In the discussion of uses a few chapters later, Grew simply fleshed out the teleology implicit in this comparison to the products of human art. The arrangement of "stalls" in the "dovecote" facilitated an efficient system for the storage of poppy seeds, while the "windows" gave those seeds necessary ventilation.[50] Grew's descriptions were integral to his broader interest in framing arguments about the design of nature. It is worth recalling, moreover, that Robert Boyle discussed the ubiquity of this style of teleological reasoning in contemporary anatomy in his *Disquisition about the Final Causes of Natural Things*, where he remarked that nothing was more common than for anatomists "to draw Arguments [. . .] about the Uses of the Parts of the Body, from their Fitness or Unfitness [. . .], to attain such Ends as are suppos'd to have been Design'd by Nature." As we saw earlier, moreover, Boyle believed that such arguments were some of the most convincing that could be made about final causes in nature.[51]

Time, Transformation, and the Ruins of Plants

The anatomical mode of description was not simply an attempt to reproduce the data of experience. The anatomical descriptions that Boyle, Grew, and Ray made were just as deeply marked by an interest in the discovery and revelation of design in nature as Hooke's representations of snowflakes. Anatomy represented a kind of empiricism in which the oft-stated desire to reproduce things as they presented themselves to the senses was balanced by an equally strong desire to find out and represent their design. We will now see, moreover, that there was an extremely close similarity between the conception of design at work in anatomical description and the account of design that informed both Hooke's interpretation of snowflakes and Inigo Jones's interpretation of Stonehenge. According to this anthropocentric and essentially static account of design, each body or structure was imagined to be the work of a designer who intended it to fulfil a specific, circumscribed set of purposes that could never change over time. Jones, for example, sought to recover what he thought to be the original Roman design of Stonehenge. His only aim in discussing the changes that the monument had undergone over the centuries since its construction was to uncover clues about its original form and the purposes that could be inferred from it. Likewise, Hooke showed little interest in discussing the process by which snowflakes came to be melted, or about what role that process might play in the economy of nature. Rather, he focused on recovering what he took to be their first design, linking that design to a specific set of purposes ordained for the formation of crystalline bodies. Both Jones and Hooke, as we have seen, privileged the recovery of a static, original design over inquiry into the dynamic processes of mutation and transformation.

To recognize that the naturalists saw design in static terms is not to deny their attentiveness to change and transformation in the bodies of plants and animals. This point is made clear by Grew's description of the emergence of "windows" in the poppy head as it dried out. He and his fellow naturalists understood that the purposiveness supposedly designed into organisms could unfold over the course of time, whether through the movements of their parts or even through the transformation of their structures into ones with new functions. Despite this interest in change and transformation, naturalists nevertheless displayed a marked tendency to interpret each structure or system as if a humanlike agent had designed it only for the fulfilment of a specific set of purposes. The account of the eye in Boyle's *Final Causes*, for example, illustrates how change and motion could be incorporated into a broader static account of design. At several points, Boyle used the eye's capacity to change shape in various ways to argue for its perfection—for instance, describing how the pupils could change shape to regulate the quantity of light that reaches the retina. While recognizing such processes of change, however, he characterized the eye as a stable system, arguing indeed that the tiniest change to its structure would make it a less effective instrument of vision than it already was. To prove the point, he referred to the many diseases that ruined the sight by bringing the eye "into a State differing from that whereto Nature had design'd it," adding a lengthy account of these at the end of the book.[52] The diseases to which the eye was liable therefore paradoxically evinced its perfection, demonstrating that changes to its structure were apt to ruin the harmonious relationship between form and function that characterized its original design. While Boyle understood that the eye was characterized by many kinds of motion and transformation, he saw these traits as preconceived aspects of a static design. There was simply no reason for a design that was already perfect to undergo transformation, and if such transformation were to occur, the result could only be a form of mutilation.

Understood as an aspect of metaphysics or theology, this static account of design in nature is well known. Writing in the 1930s, Ernst Cassirer showed that Descartes, Newton, and Leibniz all saw nature as a manifestation of divine wisdom. Consequently, they rejected the Epicurean view that nature was characterized by continual mutation caused by the chance collision of atoms. Instead, they saw nature as a static system, composed of entities that owed their essential stability to the infinite perfection of the reason that had designed them.[53] It is also well known that in the later eighteenth century, both David Hume and Immanuel Kant demonstrated that this argument was deeply anthropocentric and consequently incapable of explaining natural phenomena on their own terms. As Kant argued in the *Critique of the Power of Judgment*, one could never be certain that the procedure known by humans to be responsible for purposiveness in artificial objects—design—was in any way similar to whatever process caused nature to have the appearance of purposiveness.[54] I would like to suggest, however, that the design argument had an even deeper role in the work of naturalists such as Boyle, Grew, Hooke, and Ray than Hume and Kant perhaps realized. We have already seen that this static, anthropocentric account of design was by no means the preserve of metaphysical or theological discourse. The comparison between Jones's interpretation of Stonehenge and Hooke's interpretation of snowflakes shows instead that this conception of design was also an ideal to be pursued through practices relating to the design of both natural and artificial things. The design argument had a role to play in the material and intellectual strategies of natural history that was just as important as its role in arguments about God's relationship to creation.

In fact, naturalists often found themselves studying what we might call the ruins of plants and animals—plant and animal bodies that had undergone change or destruction over time. Moreover, the desire to recover the original designs of these phenomena was a common feature of natural-historical praxis. Some instances were so commonplace in the work of naturalists—for instance, when stuffed animals and dried plants stood in for live specimens—that they are easy to overlook. Yet naturalists evidently did not see such specimens as objects that perfectly preserved the design of plants and animals. Consider, for example, Ray's engagement with the stuffed tropicbird held in the Royal Society's repository. The sample in question possessed just two tail feathers, both of them unusually long. Ray worried, however, that this unusual arrangement was the product not of God's designing agency but rather of the conditions under which the sample had been preserved and transported. He thus wrote to Martin Lister of his suspicion that the other tail feathers had simply gone missing, adding similarly circumspect remarks in the description that he eventually included in his *Ornithology*. Troubled by the possibility that the dried sample might have been changed by the processes of its preservation and transport, Ray could not quite believe that it gave him an accurate impression of the tropicbird's original design.[55] In several other letters, meanwhile, Ray reflected on the use of dried plants, reminding his correspondents that very few of them were sufficiently well preserved to serve as dependable guides to the living plants that they stood in for. Even well-preserved samples nevertheless had to be accompanied by a great deal of contextual information about the living species to be useful.[56] Ray and his fellow naturalists regarded dried plants and animals as imperfectly preserved vestiges that, like architectural ruins, were only useful if they could be made to reveal an older, original design.

Even greater interpretative challenges arose when it came to the changes that organisms could undergo while they remained alive. As we have already seen, naturalists understood that the forms and functions of plants necessarily changed over time, meaning that many transformations could nevertheless be regarded as aspects of an original design. Yet it seemed evident that some changes were simply caused by chance events and the vicissitudes of time. A good example of the difficulties that could arise when naturalists attempted to distinguish between these two kinds of transformation can be found in a debate sparked by the 1673 publication of Nehemiah Grew's second major work on the anatomy of plants, *An idea of a phytological history propounded together with a continuation of the anatomy of vegetables, particularly prosecuted upon roots*. Here and in other works, Grew offered an account of plant growth and development with marked similarities to Hooke's hypothesis about the role of crystalline salts in the formation of more complex bodies. Grew argued that the forms of the functional parts of plants—their leaves, veins, passages, pores, and so on—were determined by the presence of the crystalline salts that could be precipitated out of the fluids contained by them. He therefore proposed that the long passages responsible for conveying sap up the root owed their forms to the similarly lengthy shapes taken by the alkaline salts that could be extracted from them. With this reasoning, Grew established an extremely close connection between form and function, arguing that the parts of plants were given form by the passage of the salts they were designed to carry. Moreover, like Hooke, he used the apparent agency of regularly figured salts to argue that there was ultimately a geometrical basis to the design of plants.[57]

The success of Grew's account of plant growth from the root therefore depended in large measure on persuading readers that the bodies he described in fact had the purposes he attributed to them. It was on this exact point that a dispute arose when, in the summer of 1673, Henry Oldenburg, secretary to the Royal Society, sought the naturalist Martin Lister's opinion of the work. As Anna Marie Roos has pointed out, Lister agreed with Grew that the veins and passages in plants had a nutritive function similar to the circulatory system in animals. Writing in early 1672, he expressed confidence that anatomical inquiries would reveal the purposes for which plants had been designed, even proposing that one of those purposes was the provision of humans with ever more powerful medications.[58] In his response, Lister nevertheless expressed doubts about Grew's chemical hypothesis, prompting an increasingly acrimonious series of letters between the two naturalists, exchanged not directly with each other but instead transmitted through Oldenburg.[59] One aspect of Lister's attack on the section dealing with the anatomy of roots is particularly striking. The problem was not just that Grew had misidentified the relationships between form and function in the parts that he had chosen to describe. He argued instead that Grew had sometimes failed entirely to distinguish between parts that were useful to the plant and those forms that had merely emerged accidentally over the course of time.

It was along these lines that Lister criticized perhaps the most spectacular image in the whole book, a double-page engraving depicting the transverse section of a vine root under magnification. Working in much the same manner as an architectural plan or section, this drawing represents the proportions of those passages, shaded in black, that Grew thought essential to the plant's development and subsequent nutrition (figure 8).

Lister's response to this magnificent image was cautiously phrased but potentially devastating all the same: "I am a little suspicious yt ye holes portrayed in ye Veine slice are merely casual intestilia or openings." In his response, Grew correctly identified Lister's remark as a threat to his chemical hypothesis. If the passages shaded in black had appeared accidentally, there would be no reason to suppose that they had a particularly important role in the workings of the plant. In turn, there would be no reason to entertain the argument that salts extracted from those same passages furnished useful insights about the plant's design. With the hint that Grew had found purposiveness in forms that had none, Lister made it possible to dismiss Grew's hypothesis as the product of an overactive imagination.[60] Grew had to respond, therefore, by reasserting the connection between form and function he had so laboriously identified, writing that everything he had chosen to include in the image was "most natural & truly organicall." The passages shaded in black were to be understood, he asserted, not as holes that had opened up by chance over the course of time but rather as organs akin to the organs of animal bodies, intended from the outset to contribute to the growth and sustenance of the plant. In other words, Grew argued that he had successfully found the original design, manifest in the relationships between form and function in the parts that he had identified.[61]

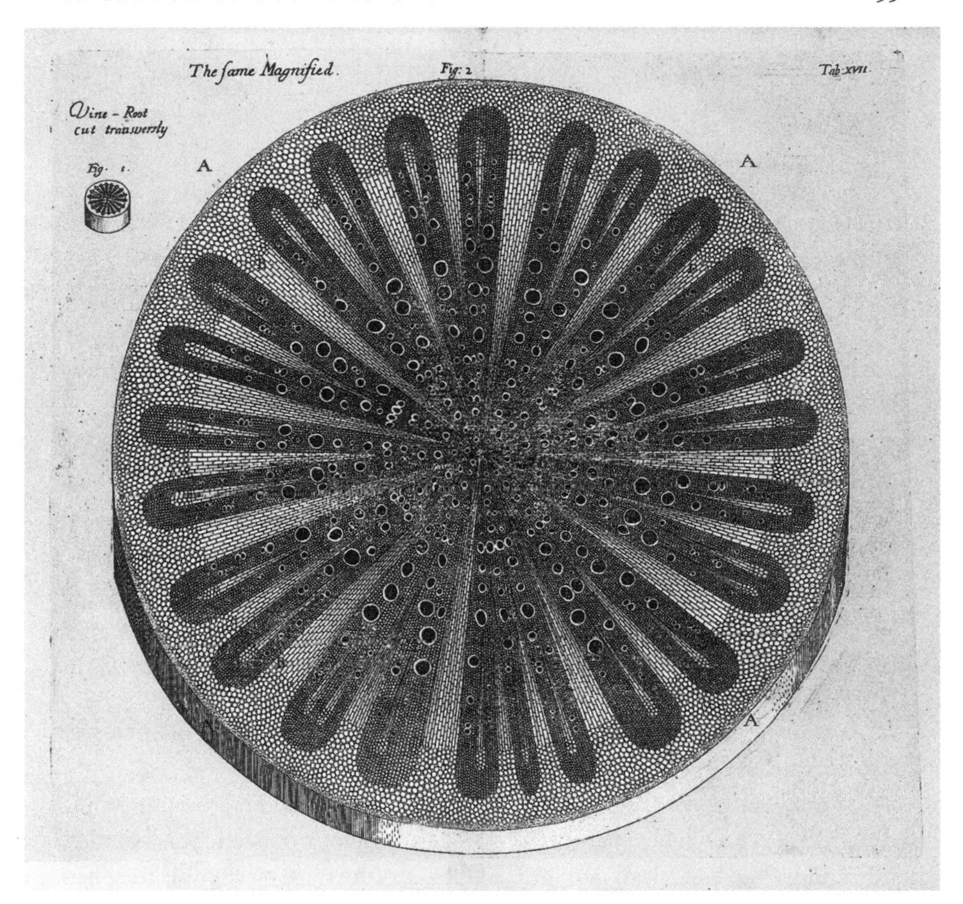

FIGURE 8. Microscopically enhanced cross-section of a vine root from the collected edition of Nehemiah Grew's works of plant anatomy, *The Anatomy of Plants* (London, 1682, Tab. XVII). (It is identical to the image originally included in the 1673 edition of the *Anatomy of vegetables*.) Wellcome Library, London. Creative Commons CC BY.

The importance of the attempt to recover a putatively original design from the undifferentiated field of empirical evidence is revealed with even more clarity in another example from the same correspondence. It can be found in Lister's response to Grew's account of the parenchymous tissues found in in the pith of the root and stem, along with the underside of the bark. Taking inspiration from Hooke's observations on the texture of cork in the *Micrographia*, Grew had described them as being composed of countless tiny bubble-like structures.[62] He gave those bubbles an important role in his chemical hypothesis, seeing them as the plant's instrument for fermenting, and extracting nutrients from, the sap. The passage of the sap from bubble to bubble on its journey from the root to the leaves was to be the mechanism both for it to be endlessly refined and for the saline particles to pass into whichever part of the plant happened to need them.[63] Of course, Lister found this description unconvincing, offering as counterexamples a number of plants that appeared to have pith of a rather different conformation, such as the walnut tree.[64] Grew's response was not to deny what Lister had seen but instead to question its significance, much as Lister had done with the passages in the vine slice. Grew admitted that the pith sometimes appeared to have a shape other than the one he had given it, even adding a few examples of his own to those supplied by Lister. For Grew, however, these forms were merely deviations from the plant's intended design, explaining them away as "nothing else but so many ruptures, and by no meanes appertaining to the original conformation of the pith." Lister had, in Grew's view, mistakenly identified as aspects of the plant's intended design forms that had emerged accidentally. Grew went on to point out that the intended design could be deduced from observations made on the pith in "the most young & succulent parts" of the plant, where the original bubble-like structures were more readily visible.[65] In this way Grew asserted that the original design should take precedence over other forms in the hierarchy of representation. He knew that the pith was apt to change shape over time, but he believed that the bubble-like structure observable in the young plant was nevertheless the structure that most closely reflected the designer's intentions.

We can see now see that Grew, like Jones and Hooke before him, subordinated his representational praxis to a static and anthropocentric account of how things had been designed. Indeed, he used the same strategy as they had when confronted by evidence that appeared to contradict his view of natural things as the geometrically composed material instantiations of divine purposes. As in the case of Hooke's snowflakes, he characterized the aberrant forms as ruins. They were not to be understood as evidence against design but instead as clues about the existence of designs that had once exhibited all the usual symptoms of divine excellence. Recall that Lister expressed a similar attitude toward such forms when he castigated Grew for focusing so much attention on the possibly purposeless "casual interstilia or openings" in his image of the vine slice. As we know, Grew was also capable of seeing changes in the forms of plant and animal bodies as aspects of an original design. Yet his attempts to integrate such changes into his otherwise static account of design produced difficulties that he was not capable of resolving. When, for instance, he revised his earlier works of plant anatomy for the 1682 edition of *The Anatomy of Plants*, he added a brief acknowledgment of the pith's propensity to change shape, citing the same examples he had given to Lister in the preceding decade. This time, however, he did not dismiss the aberrant forms as ruins. Instead he made a weak effort to portray them as products of design, highlighting their regularity and asserting that they "always" emerged "for good use," albeit without offering any suggestions about what that use might be. He did nothing, moreover, to square this admission of change over time with his broader description of the anatomy of plants. Just one page later he unfolded a completely static metaphor to illustrate the fundamental structure of all plants in existence, likening the bubbles in the pith to the tiny holes observable in a piece of bobbin lace pinned into place on a cushion. Grew did not let the knowledge that the pith could sometimes change shape trouble his image of plants as perfect, static pieces of divine needlework.[66]

An Empiricism of Beauty and Utility

For Grew, uncovering design in the bodies of plants and animals transformed the otherwise banal work of anatomical investigation into a form of moral and theological edification. To discover perfect correspondences between form and function in the pores and passages of plants was not simply to learn about how plants worked. It was to produce an endlessly detailed revelation of the operations of God's "Universal Monarchy" in the material world. Grew declared at the very start of his account of the vegetation of roots that the encounter with final causes was what made plant anatomy worthy work for a Christian philosopher. "To Philosophize," he wrote, "is, To render the *Causes* and *Ends* of Things. No man, therefore, that denieth *God* can do this, Truly."[67] Such assertions about the meaning and ultimate purposes of natural history had an obvious apologetic function. However, it is clear from the examples we have encountered that those assertions found an equally important expression in the practices through which Grew and others in the Royal Society sought to produce knowledge. When Boyle, Grew, Hooke, and Ray inquired into and represented natural things, they worked in a way that reflected their belief that God had designed them. Since God was in some respects similar to humans, it was legitimate to interpret some of God's works with the resources afforded by architecture and other human design practices. At the same time, however, the necessary perfection of God's designing agency made the appearance of imperfection troubling. To preserve their image of divine perfection, naturalists interpreted imperfect phenomena as the ruins of forms that had once exhibited beauty and purposiveness in the highest degree.

Lorraine Daston and Peter Galison identify a similar tension in their account of scientific objectivity in the seventeenth and eighteenth centuries. On one side, naturalists wanted to obtain accuracy by mechanically reproducing the data of experience, even advocating the use of such instruments as the camera obscura for that purpose. On the other, they generally believed that the best way to represent natural kinds was by turning to putatively "perfect" or "characteristic" types. Until the nineteenth century, Daston and Galison therefore argue, the conspicuous exercise of philosophical judgment remained essential to the image of scientific objectivity. Only by exercising this judgment could the philosopher identify, or in some cases even produce, the perfect specimens required for representing the species of natural things.[68] The naturalists of the early Royal Society shared, in its broad outlines, this premodern image of objectivity, seeing the exercise of judgment in the discovery of perfect types as necessary to the production of scientific representations. However, little has been said about how naturalists went about judging the perfection of individual specimens. Indeed, their ultimate reasons for regarding perfect specimens as particularly instructive have often remained in the dark. The society's engagement with ruins enables us to answer these questions. The criteria for judging the perfection of nature were the criteria of human design practices. Natural things were judged to be ruins when they failed to meet the standards of beauty and purposiveness to be expected from examples of good design. The task at hand was not the representation of things in their totality but instead the production of representations that granted access to the ideas that came prior to those things—their designs. It was only by recovering perfect designs from the otherwise undifferentiated field of sensory experience that naturalists could produce encounters with the original, intended meanings of nature. The exemplarity and instructiveness of perfect specimens came from their capacity to communicate the design in which their true purposes, and consequently true meaning, had been encoded.

Recognizing the importance of design practices makes it possible, moreover, to identify a hitherto unremarked connection between architecture and natural history in the work of the early Royal Society. They were linked by their shared concern with interpreting the products of design, both at the conceptual level and at the level of practices. This is not to suggest that naturalists thought that God worked exactly like an architect or that they interpreted natural things as if they were buildings. Instead, as the comparison between Jones's treatment of Stonehenge and Hooke's treatment of snowflakes reveals, natural history shared with architecture the premise that its objects were the products of design, along with a sense of what the products of design were like. Whether natural or artificial, well-designed objects were thought to comprise a harmonious combination of utility and beauty. This ideal, familiar to us now from Hooke and Grew, was one of the founding tenets of architectural theory. In the first of his "Tracts" on architecture, Christopher Wren defined the "Principles of Architecture" as "Beauty, Firmness, and Convenience," translating Vitruvius's famous triad of *venustas*, *firmitas*, and *utilitas*. As Wren's use of Vitruvius indicates, there was nothing fundamentally new in the simple assertion that beauty should be a corollary of good design. J. A. Bennett has shown, however, that Wren gave the old claim a new theoretical basis—one that reflected his dual commitment to empiricism and the mathematical sciences. Rather than define beauty as a property that actually belonged to any form in particular, he instead described it as the experiential product of underlying natural principles that could be discovered through observation. For Wren, these "natural Causes of Beauty" were regular geometrical principles—principles to which the forms and functions of natural phenomena could ultimately be reduced. Positioning architecture as an artificial expression of such principles, Wren made it possible to identify architectural beauty as an effect caused naturally by good design.[69]

As we have seen, Grew and Hooke shared Wren's understanding of beauty as an effect of the geometrical order to be found in natural phenomena. They both aimed to prove that nature's purposes depended on regular, geometrical arrangements of crystalline particles. Yet the comparison between architecture and natural history becomes even more instructive if we turn to Wren's engagement with ruins. One of the ruins that interested him most was that of a vast structure in Rome known then as the Temple of Peace but now understood to be the much later Basilica of Maxentius and Constantine. After briefly describing the building in its original state, Wren suggested that the design did not in fact measure up to his standards of beauty. The focus of his criticism was the building's rather flat appearance, an effect produced by the relatively small size of its portico as well as the great breadth of its wings. At this point, however, Wren abruptly changed course, explaining that these seemingly strange design choices could be explained by the temple's purpose—its dedication to peace. Moreover, he argued that understanding the design as an expression of this purpose enabled one to see that the structure was actually very beautiful:

> It was not therefore Unskilfulness in the Architect that made him chuse this flat kind of Aspect for his Temple, it was his Wit and Judgment. [. . .] No Language, no Poetry can so describe Peace, and the Effects of it in Men's Minds, as the Design of this Temple naturally paints it, without any Affectation of the Allegory. It is easy of Access, and open, carries an humble Front, but embraces wide, is luminous and pleasant, and content with an Height it might otherwise justly boast of, but rather fortifying itself on every Side, rests secure on a square and ample Basis.[70]

Wren's analysis of the Temple of Peace shows us that he did not simply see architectural design as a compromise between the competing demands of beauty and utility. Instead, he identified beauty as a necessary experiential outcome of the discovery of purposiveness in design. In this respect, his attitude to the temple's aesthetic value closely resembles Grew's fluctuating attitude to the structure of the pith in the roots and stems of plants. Where Grew saw irregular forms, he also saw forms without purpose. When, however, he came to suspect that forms in fact had a purpose, he tried to suggest that they were beautiful after all. We can therefore see that in these instances, architectural criticism and natural history shared the same general approach to interpreting the products of design. The empirical discovery of utility in natural and artificial bodies led necessarily, they thought, to an experience of beauty.

In this chapter, we have encountered several suggestions that concerns about the experience of beauty and sensory pleasure—concerns that would today fall under the rubric of aesthetic experience—had an important role in the empirical sciences. Once again, a comparison with Wren's work on architectural design is illuminating. Recall that Hooke took a question about the experience of beauty as the *Micrographia*'s premise, framing the work as an instrument for overcoming the sensory corruption that prevented people from experiencing nature in the way God originally intended. Wren was equally troubled by the fact that people did not always experience beauty when he thought they should. His response, if less explicitly theological than Hooke's, was otherwise very similar. He placed the blame squarely with the observer. An ugly building might very well induce an experience of beauty, but only if an observer's judgment had been corrupted by habitual misuse of the senses: "Familiarity or particular Inclination breeds a Love to Things not in themselves lovely. Here lies the great Occasion of Errors."[71] Wren reasoned that architectural history was the most appropriate means of remedying such defects of judgment. The encounter with persuasive representations of the best examples of ancient architecture would, he hoped, communicate both an appreciation of the principles of design and a capacity to experience beauty when it was appropriate. Thus he positioned the fifth of his architectural treatises as an "endeavour to reform the Generality to a truer tast in Architecture by giving a larger Idea of the whole Art, beginning with the reasons and progress of it from the most remote Antiquity."[72] In Wren's suggestion that learning the "reasons" of architecture would lead to the acquisition of a "truer tast," there is another hint that he linked the process of gaining knowledge to the development of a capacity for appropriate kinds of aesthetic experience. The same hint is echoed in Hooke's dramatic presentation of the *Micrographia*'s revelation of nature's fundamental principles as an exercise that would give readers the capacity to find sensory gratification in objects that had once seemed unpleasant to behold.

We might therefore say that Hooke and Wren saw their work as an *aesthetic science*. Theirs was a form of empiricism in which the use of the senses in the pursuit of knowledge was to be accompanied by the experience of beauty or sensory pleasure. In the remaining two chapters, I will show that this interest in aesthetic experience played an important role in a wide range of contemporary anatomical and natural-historical works. Naturalists such as John Ray and Thomas Willis believed that the capacity to experience appropriate forms of sensory pleasure—the capacity for a particular kind of subjectivity—betokened the ability to participate in the work of empirical knowledge production. They therefore used representations of nature not simply to communicate about the designs of natural things but indeed to communicate the capacity to reckon with those things in an appropriate manner. The aim of natural history was not simply the transmission of knowledge. Rather, it was the communication of the affective disposition necessary to gain intellectual, moral, and theological benefits from the study of natural phenomena. To understand how natural history was thought to produce such wide-ranging effects, it is, however, necessary to completely reassess the medium of representation that was then most widely used. Few people would dispute that Hooke wanted to produce a powerful affective response with the dazzling pictorial representations of the *Micrographia*.[73] However, the majority of contemporary natural-historical and anatomical representations were not pictorial at all—they were verbal descriptions. The prevailing view of those verbal descriptions is that they were written in a plain style, intended to transparently signify the objects that they stood in for. We will see, however, that they were in fact highly rhetorical, intended not just to represent the world but also to produce powerful affective and aesthetic experiences in the minds of readers.

4

Verbal Picturing

John Ray hoped that his greatest work, the *Historia Plantarum* (1686–1704), would be published with illustrations.[1] During the 1680s and 1690s Ray discussed the possibility of including engraved copperplate pictures with several of his confidants. Among them was Tancred Robinson (1657/8–1748), a physician and naturalist who had recently become one of Ray's most important supporters. Around the second half of 1684, in a letter that does not seem to have survived to the present day, Robinson advised Ray not to print the *Historia Plantarum* with pictures, seemingly because they would make the book too expensive to be of use to its readers. In his reply of October 1684, Ray explained that "others" disagreed entirely with Robinson's advice. They had, as Ray put it, warned him that a history of plants without pictures would be as deficient as an atlas without maps. Confessing that he agreed with this assessment, Ray added that he saw pictorial representation as far superior to the verbal descriptions with which he would otherwise have to make do:

> A good Figure conveys that to the Mind suddenly, and with Ease and Pleasure; an Idea whereof cannot be formed by the Help of a Description without Time and Pains, and a greater Attention than most Readers have Patience to give it.[2]

This brief assertion of the superiority of pictures over words is striking for its emphasis on the affective dimensions of visual experience. Ray asserted, indeed, that the success of a natural-historical representation was to be determined by its capacity to provoke certain kinds of apparently subjective experience. Those experiences were, as he explained, the suddenness, ease, and pleasure attendant upon visual experience. The power of pictures came from their capacity to arouse pleasurable feelings. Ray valued pictorial representations because they could induce those feelings to a far greater extent than verbal representations could.

Ray's assertion of the superiority of pictures over words presents us, however, with an interpretative challenge. A close reading of the passage suggests that Ray did not distinguish much between the final mental ideas that could result from these two distinct forms of representation. Instead, he asserted that pictures could lead the mind to form those ideas more quickly, and with more "Ease and Pleasure," than could be accomplished with words. He did not suggest that pictures led the mind to form images that were better, for example, by more closely resembling the things for which they stood. His assertion of the power of pictures was based not on an assessment of their capacity to lead the mind to ideas that better resembled their objects than words, but rather on an assessment of their affective power. Ray's interest in the affective power of pictures leads, however, to an apparent paradox. In this chapter, we will see that Ray drew on famous ideals of *verbal* representation when he asserted the superiority of pictures. His emphasis on the vivid, easy, and pleasurable mental picturing attendant on visual experience had its sources in rhetorical theory—especially a rhetorical ideal of vivid description known as *enargeia*. In the early modern period, rhetoric amounted to a kind of psychological theory, detailing the ways in which orators could persuade audiences by provoking the passions. Above all else, rhetoric valued representations that were capable of commanding assent because of their capacity to effect transformations in the affective states of listeners and readers. In drawing on the ideal of *enargeia*, Ray was therefore also drawing on the passionate, affective model of persuasion that had long sustained rhetorical forms of speech and writing.

In this chapter and the next, we shall investigate the interplay between words and pictures in the strategies for representing nature used by Ray and some of his key contemporaries, including Nehemiah Grew, Robert Hooke, and Thomas Willis. The aim of this investigation is to recover the rhetorical and pictorial qualities of their supposedly plain, passionless verbal descriptions. The plain style of description, so often held up by the Royal Society of London as the model for objective communication, was in fact intended to induce feelings of pleasure just like those that Ray attributed to visual experience. The point therefore is not merely to show that naturalists employed rhetorical strategies in the work of verbal description, or even that they drew on rhetorical models when talking about the use of pictures. Instead, we will see that they believed it possible to use words to reproduce the experience of vivid picturing, or even the experience of visually encountering things themselves. Mobilizing the rhetorical strategies associated with *enargeia*, moreover, they held up the possibility of reproducing, with words, the pleasures and pains that often accompany such an experience. Ray and his fellow naturalists, therefore, did not turn to *enargeia* simply because of its importance in rhetorical models of vivid and persuasive description, or even because of its importance in contemporary theories of art. They did so because the ideal of *enargeia* both reflected and shaped the conception of experience that flowed from their ideas about the mind and its mechanisms. Indeed, key texts on the workings of the mind such as John Locke's *Essay Concerning Human Understanding* and Willis's *Cerebri Anatome* enable us to understand how and why Ray and his contemporaries turned to the literary strategies associated with *enargeia*. Drawing these texts into our discussion, it will thus be possible to reconsider the place of the affective dimensions of experience in their approaches to the communication of knowledge.

This reconsideration is crucial because most of the representations produced in the Royal Society's ambit were verbal descriptions. Any interpretation of the representational practices of seventeenth-century natural history and anatomy must reckon with them and attempt to integrate the insights gained with those gleaned from the study of representations in other media. A good illustration of this point is to be had from the fate of the *Historia Plantarum*, the book that occasioned Ray's remarks on the power of visual representation. Notwithstanding Ray's assertion of the superiority of pictures, the work was eventually published without the hoped-for illustrations. It is likely that the commercial failure of Ray and Francis Willughby's illustrated history of fish, *Historia Piscium* (1686), made it impossible to raise the enormous sum of money that would have been needed to have the engravings made.[3] The case of the *Historia Plantarum* serves as a stark reminder that Ray and his fellow naturalists were, in most cases, forced to rely on verbal descriptions to represent the forms and textures of natural things. Even in heavily illustrated books, such as the *Historia Piscium*, a very large part of the work of representation was in fact assigned to the verbal descriptions. This was not, however, only for reasons of cost. As this chapter will show, naturalists had powerful philosophical reasons for seeking to represent some things by verbal, rather than pictorial, means. It was possible, they thought, to use descriptions to reprise or somehow reproduce the mental effects of sensory experience—even the feelings of pleasure that Ray attributed to vivid picturing.

To rediscover the sensory qualities of these descriptions is, therefore, to fundamentally reconsider their value as evidence about the empirical project pursued by Ray and his contemporaries. In this chapter, we will see that Ray's descriptions were intended to produce just the kind of ease and pleasure that, as we have seen, he attributed to pictures. Through a sustained stylistic analysis, meanwhile, chapter 5 will show that the evocation of sensory pleasure was central to the knowledge-producing work performed by those descriptions, and therefore to the epistemological project of which they were the most important part. Ray and his contemporaries sought to shape the way in which their readers experienced natural-historical evidence, but they did not seek, as is most often claimed, simply to eliminate pleasure and the other subjectivities of experience. Instead, they wanted to engender in their readers a specific kind of subjectivity—a pleasurable affect deemed to be consistent with, and even helpful to, the mental exercise of imagining natural things.

John Ray, the Royal Society, and the Pleasures of Rhetoric

Roughly six years after he wrote to Tancred Robinson about the power of pictures, John Ray sent him a second letter touching on rhetoric. Composed in December 1690, this letter was a response to Robinson's proposal for a book comparing the achievements of ancient and modern learning. In a departure from his usual tendency to assess questions evenhandedly, Ray responded by bluntly asserting the superiority or parity of modern learning in every area except one. The sole exception was the art of rhetoric, which Ray referred to as "Acuteness of Wit and Elegancy of Language." Ray argued, however, that the skill of the ancients in the rhetorical arts was by no means a compelling argument in their favor. Rhetoric, he pointed out, had little value in comparison to the studies at which the moderns excelled, such as natural history and experimental natural philosophy. To reinforce the point, he made a brief attack on rhetoric, arguing that it consisted in nothing more than encouraging readers to respond passionately to things that had no true worth:

> But those [rhetorical] Arts are by wise Men censured as far inferior to the Study of Things; Words being but the Pictures of things, and to be wholly occupied about them, is to fall in love with a Picture, and neglect the Life.[4]

Ray thus asserted that rhetorical forms of expression could induce a singularly powerful passion—"love." As powerful as this passion was, it was nevertheless misdirected. Rhetoric could stoke strong passions, but it did so through objects that had little use or value in themselves. It was better, Ray explained, to be preoccupied about solid things than mere words.

Comparing this attack on rhetoric with Ray's previous remarks on the power of pictures, an apparent contradiction arises. His account of the power of pictorial representations endorsed precisely the affective model of persuasion that he went on to attack in his letter of 1690 on ancient and modern learning. When Ray discussed the power of pictures, he saw their capacity to induce pleasurable feelings as a virtue. When he discussed rhetoric, however, he identified the same quality—or at least one very similar—as a bad thing. How can we account for this inconsistency? Ray does not appear to have changed his mind about the role of affect in pictorial representation over the intervening six years. In fact, he repeated his claims about the power of picturing almost word for word in a 1691 review, for *Philosophical Transactions*, of the illustrated botanical work *Phytographia, sive Stirpium Illustriorum* (London, 1691–1705) by Leonard Plukenet (ca. 1642–1706).[5] Another possibility is that Ray drew a sharp distinction between pictures and words, seeing the affective power of the former as useful to the production of knowledge and harmful in the case of the latter. The problem with this supposition is that, as we have already seen, Ray and his contemporaries did not in fact distinguish clearly between the effects that could result from words and pictures. This lack of distinction is again evident in Ray's attack on rhetoric, which refers to words, confusingly from today's vantage point, as "the Pictures of things."

To understand the meaning of Ray's attack on rhetoric, we must place his stated views in context. Ray's dismissal of rhetoric closely followed the Royal Society's corporate critique of rhetoric—one that it had been pushing since its earliest days. The most famous expression of this critique is to be found in Thomas Sprat's *History of the Royal Society* (1667), a defense of the society's activities that came close to expressing its corporate position on a number of issues.[6] In that text, Sprat assailed rhetoric in terms much like those Ray would use in his letter to Robinson some twenty-five years later. It was the appeal of rhetoric to the passions that made it so deceptive. By privileging the pleasures of figurative speech over the accurate use of words, rhetoric destabilized the relationships between words and the ideas for which they stood. Through its pleasurable distortions, rhetoric could be used to motivate people to accept and even act on falsehoods, overcoming the scruples of reason. In other words, the pleasures of rhetoric concealed an insidious purpose. In this connection, Sprat remarked that rhetoric could even be used to accomplish the perverse end of giving people a taste for falsehood over things that were both true and morally improving. As he put it, the pleasures of rhetoric were too often used to "make the Fancy disgust the best things, if they come sound, and unadorn'd."[7]

Through Sprat's *History* and several other works, members of the Royal Society thus mounted what appears to have been a fierce attack on rhetoric, insisting that its emphasis on the inducement of pleasure and the destabilizing effect that this emphasis had on language made it an impediment to truthful communication. This attack was accompanied by a now equally famous set of attempts to show that the society had found a better way to communicate. In contrast to rhetoric, the language used by the Royal Society supposedly comprised a transparent and neutral form of communication. It could, the society asserted, serve to objectively represent objects of experience as they presented themselves to the senses. As Sprat explained, this new form of communication consisted in using an entirely plain style of description, devoid of all the pleasure-inducing rhetorical ornaments that could lead the mind astray. In choosing the plain style, the Royal Society had, he wrote, chosen "to return to the primitive purity, and shortness, when men deliver'd so many *things*, almost in an equal number of *words*."[8] Sprat thus claimed that the society's words had an unambiguous relationship to the conceptions for which they stood, reflecting in some measure the perfection of the language that he and many contemporaries took to have been in use either before Adam's ejection from the Garden of Eden or God's destruction of the Tower of Babel.[9] According to the *History*, the use of the plain style would do more than produce clearer knowledge of natural phenomena. It would, like Hooke's microscope, assist in the restoration of humanity to a more virtuous condition.[10] The Royal Society's ambitious stylistic program has long occupied an important place in the history of language theories, along with accounts of the emergence of modern forms of scientific representation and the rise of modern literary aesthetics. It has often been argued, moreover, that the use of the plain style was a vital part of the society's program for disciplining the passions in its pursuit of objectivity. Through its attack on rhetoric and self-conscious use of a plain style, the Royal Society apparently ruled out the possibility that affectively powerful forms of representation could play a useful role in the communication of natural knowledge.[11]

But it has been a long time since modern scholars have taken the Royal Society's disavowals of rhetoric entirely at face value.[12] In the early 1980s, for example, Steven Shapin uncovered the rhetorical strategies of Robert Boyle's famous descriptions of experiments with the air pump, published in 1660 as *New Experiments Physico-Mechanicall, Touching the Spring of the Air* (1660). Boyle's apparently plain style concealed the deployment of a variety of rhetorical strategies intended to produce an *effect* of neutral and transparent representation. Among the most important of those strategies were the use of dense descriptions intended to situate individual experiments in a specific time and place, and the literary production of a modest authorial persona, consistent with the straightforward narration of events as they took place.[13]

More recently, scholars have built upon Shapin's demonstration by showing that key figures of seventeenth-century philosophy continued to take a keen interest in affective powers that had long been attributed to the pleasures of rhetoric. While philosophers worried about the possibility of deception, they admired rhetoric for its capacity to give readers clear and distinct ideas that might otherwise have been much harder for them to grasp. Quentin Skinner has shown, to take an influential example, that in roughly the second half of his career, Thomas Hobbes rejected his own earlier attempts to eliminate verbal figuration from philosophical argument. Instead, he embraced the persuasive psychology of rhetoric, filling his famous work of political theory, *Leviathan* (1651), with a host of rhetorical devices designed to make his claims more memorable and persuasive. Not least among those devices was the title of the work itself, likening his ideal commonwealth to the vast sea monster of the Old Testament.[14]

Even the position articulated by Sprat in the *History* was more nuanced than it seems. In a passage about three hundred pages after the well-known attack on rhetoric, Sprat argued that the Royal Society's plain style could form the basis of a reformed, virtuous rhetoric. This rhetoric would be full of vivid and pleasing images derived from descriptions of nature and its operations.[15] Thus he seemingly suggested that the pleasures of rhetoric, once allied to truly useful and virtuous objects, could be turned into instruments for the vivid, persuasive communication of the truth. While Sprat only hinted at this possibility, there is much evidence to show that other members of the Royal Society put it into action.[16] Lawrence M. Principe has shown that Robert Boyle was deeply disturbed by the power that French romances—works by La Calprenède, Montauban, and Corneille—had held over his mind during his youth. The experiences of the pleasure he derived from reading those romances led him to deploy literary strategies derived from them in his own works. Boyle hoped in this way to incite readers to take a similar pleasure in works that would contribute to their store of knowledge and improve their moral disposition—moralizing tracts and works of natural philosophy. The pleasures of rhetoric were to be harnessed, or so Boyle hoped, to the cause of making otherwise unpalatable moral and natural-philosophical discourses as pleasurable as he had once found corrupting romances.[17] The Royal Society's attacks on rhetoric therefore concealed a far more contradictory attitude to the use of the affective strategies of rhetoric and poetry. Despite their own warnings about the possibility of deception, several key members of the society made use of rhetorical forms of expression in their works. They hoped, by inducing experiences of ease and pleasure, to give readers clear and persuasive ideas of the things and notions that they described.

Picturing Natural Kinds

We can now partially resolve the tension between John Ray's praise for the affective power of pictures and his suspicion of the same property in rhetoric. His attack on "Elegancy of Language" reflected the Royal Society's corporate position on rhetoric, a position that concealed a continuing interest in the strategies of rhetorical persuasion. Ray in fact equaled his colleagues in the use of figurative language, filling his natural histories, as we shall soon see, with vivid descriptions, laden with metaphors. We are left, however, with the task of understanding why Ray and his fellow naturalists believed that pictures were more powerful tools of representation than words. It might be assumed that he and his contemporaries premised the superiority of pictures on their mimetic power—their apparently superior ability to lead the mind to ideas truly resembling their objects. We have seen, after all, that it was the potential inaccuracy of metaphors and other figures of speech—their unstable connection to their referents—that made rhetoric so unreliable. It would be mistaken, however, to assume that early modern naturalists and natural philosophers took the mimetic power of images for granted. Indeed, Sachiko Kusukawa and Claudia Swan each have shown that the use of pictorial strategies provoked controversy among naturalists of the sixteenth century. The naturalist and physician Leonhart Fuchs (1501–1566), for instance, justified the use of images in natural history by arguing that natural kinds could reliably be identified through the pictorial representation of their external features. Yet Fuchs's contemporaries were by no means universally convinced. The physician Sébastien de Monteux (1518–1559) responded, for instance, by using Aristotelian arguments to cast doubt on the notion that plants could be identified by their external features alone. About a decade later, moreover, the physician Janus Cornarius (1500–1558) objected that pictures could never teach people to distinguish between plants unless they had first seen the plants with their own eyes. In the sixteenth century, therefore, the notion that images could reliably stand in for real things was a matter of controversy, implicated in debates about the nature of experience and its role in leading us to distinguish between one species and another.[18]

In the latter part of the seventeenth century, naturalists addressed similar questions—and mobilized many of the same ancient concepts and sources—in their own debates about how to use pictures in the representation and identification of natural kinds. Ray was impressed, as we have seen, by the power of images. Nevertheless, like some of Fuchs's critics, he was unconvinced that pictures alone could teach people how to distinguish between species. We can get a better sense of Ray's position by considering his use of pictures in a little more detail, and by comparing his approach with that used by some of his immediate contemporaries. When Ray represented natural kinds, he either used words alone, as in the *Historia Plantarum*, or he used words and pictures in combination, as in his *De Historia Piscium*. This latter work is a heavily illustrated history of marine life, which Ray worked on with his patron Francis Willughby until the latter's death in 1672.[19] With its combination of words and pictures, the *Historia Piscium* resembles most of the natural histories then printed with illustrations. It differs sharply, however, from two contemporary works—Martin Lister's *Historiae sive Synopsis Methodicae Conchyliorum* (1685–1692), and Leonard Plukenet's *Phytographia*.[20] In those works Lister and Plukenet pursued a more thoroughly pictorial approach, accompanying images of species with very little text at all. Comparing Ray's mixed-media approach to representing species with the pictorialism of these two works, we may see that Ray had far less faith than Lister and Plukenet did in the mimetic power of pictures.

Both Lister's *Historiae Conchyliorum* and Plukenet's *Phytographia* depend almost exclusively on pictures to do the work of identifying natural kinds. Lister's work details the species of creatures with shells, using beautiful illustrations realized by Lister together with his talented young daughters. Meanwhile, Plukenet's book represents plant species.[21] Both books, moreover, were produced in a way that established the pictures as by far their most important component. Neither book contains any typeset text whatsoever. Instead, each page was printed from engraved copper plates, and nearly every one of these contained large pictures depicting the species of shelled creatures or plants, respectively. The small quantity of text was realized through intaglio engraving onto the plates themselves. As a result, the pictures in each book are accompanied by a small amount of text rendered in a pretty cursive script (figures 9 and 10).

There are some stylistic and technical differences between the pictures in these two books. Lister's shells are both more skillfully realized and lifelike, while Plukenet's plants are stiff and flat in appearance, perhaps indicating that the models for the drawings were dried, flattened samples from an herbarium.[22] Nevertheless, the two naturalists pursued roughly the same approach to visualizing natural kinds. This approach was genuinely pictorial, depending in large part on illustrations of the external parts, respectively, of shells and plants. The text played a small role, for the most part simply itemizing the things pictured in the plates.

This pictorial approach was underwritten, I think, by two interconnected assertions about the role of visual experience in natural history. Both Lister and Plukenet hoped, first, that their pictures would, through a combination of their beauty, their cost, and the quality of their execution, cause pleasure. As

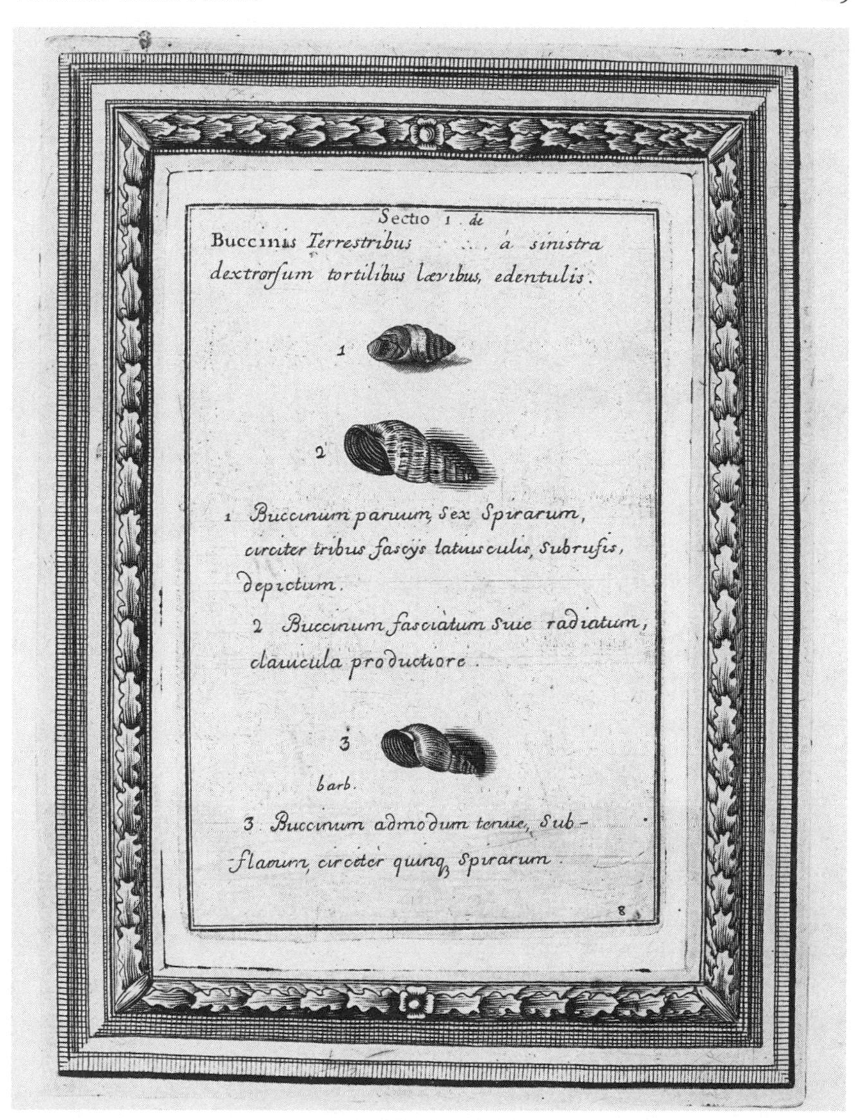

FIGURE 9. Engraved folio plate showing mollusk shells, from Martin Lister, *Historiae sive Synopsis Methodicae Conchyliorum*, vol. 1 (London, 1685), plate 8. By kind permission of the Syndics of Cambridge University Library.

Anna Marie Roos has shown, Lister invested enormous resources of money, materials, and skill, along with the artistic talent of his daughters, to ensure that his plates would be supremely beautiful. He even went to the trouble of encompassing some of the illustrated shells in ornate borders (see figure 9), produced by engraving the borders onto separate plates and separately adding them to the page.[23] The intended effect of Plukenet's plates, meanwhile,

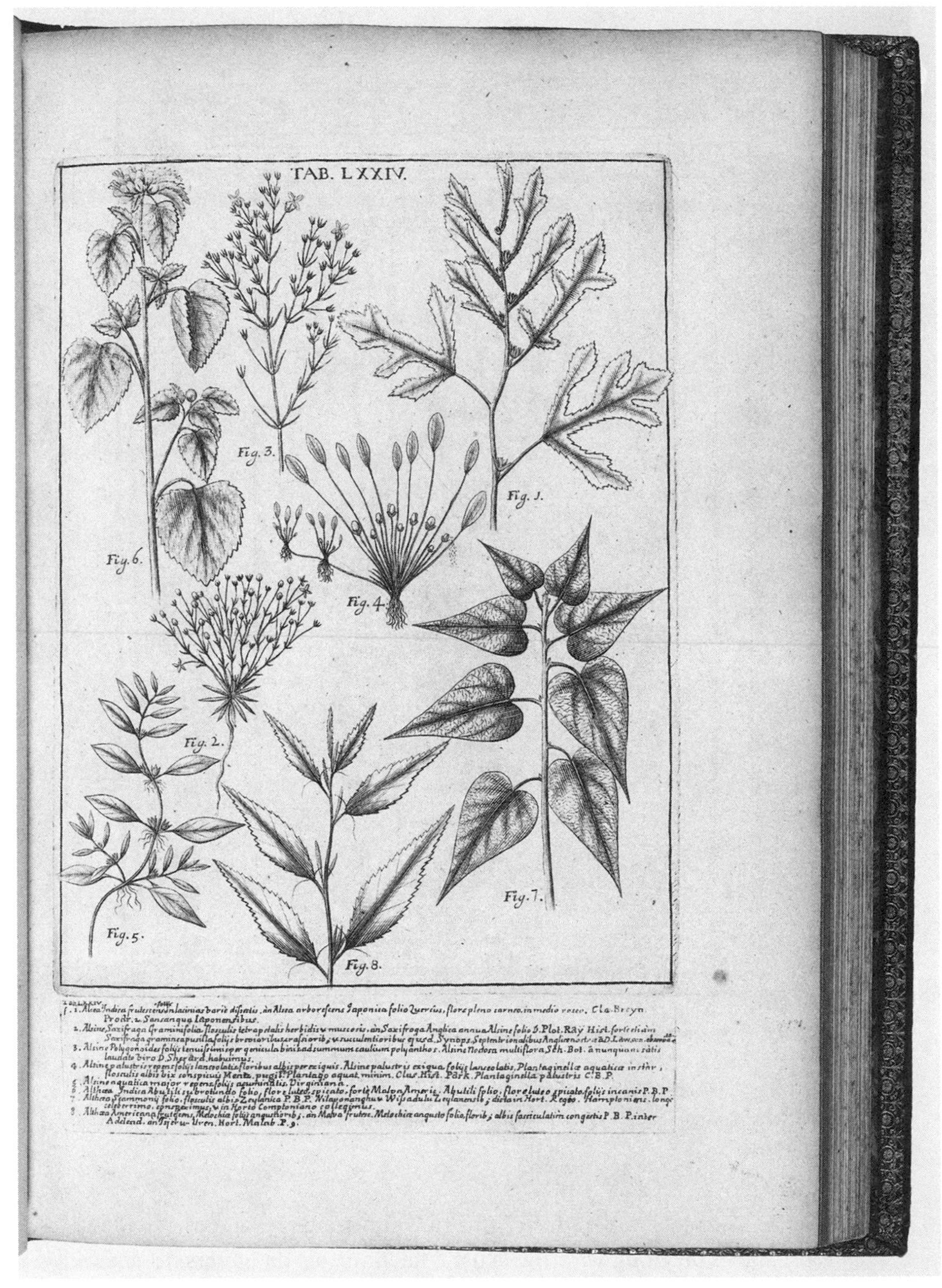

FIGURE 10. Engraved folio plate showing various plants, from Leonard Plukenet, *Phytographia, sive Stirpium Illustriorum*, vol. 1 (London, 1691), Tab. LXXIV. By kind permission of the Syndics of Cambridge University Library.

may perhaps be gathered from Ray's dutiful review of the *Phytographia* in *Philosophical Transactions*—the one in which he repeated his assertions about the power of pictures. There, Ray remarked that the "beauty & elegancy" of Plukenet's pictures would win them favor with a broad audience.[24]

Furthermore, Lister's and Plukenet's insistence on the beauty of the individual plates dovetailed with a second assertion about how a specific form of visual experience or practice could facilitate the identification and distinction of natural kinds. In both books, Lister and Plukenet repeatedly made use of the same strategy for visualizing species and the differences among them. As we can see in figures 9 and 10, this strategy comprised the pictorial representation of the exterior parts of shells and plants—those parts that are visible to the naked eye in one view. Indeed, the burden of representing species in both the *Historiae* and *Phytographia* lay on the engraved pictures of the exterior parts of whole shells, whole plants, or large parts of plants such as entire branches.[25] Lister and Plukenet thus tacitly asserted that it was for the most part possible to distinguish the species of shells and plants by comparing their visible, external surfaces.[26]

Although this boldly pictorial strategy seems to correspond closely with Ray's own ideas about the ability of pictures to cause clear, vivid mental images, Ray never used it himself. Instead, as we have seen, he consistently relied either on a mixture of verbal and pictorial representations or on verbal descriptions alone. Ray's reasons for doing so ultimately sprang from a position, shared by many of his colleagues in the Royal Society, about the nature of sensory experience and the status of the inferences that could be derived from it. Among Ray's works, the one that most closely resembles Lister's *Historiae* is the *Historia Piscium*, since it contains 189 plates depicting the species of fish. These plates, moreover, have a similar general appearance to those in Lister's book. They represent entire fish viewed head to tail from the side, either individually or in groups of three. There are just a few exceptions to this rule, including the pictures of kinds of rays, which depict them either dorsally or ventrally because of their flat profile. The key difference between the *Historia Piscium* and Lister's *Historiae*, however, is the presence in Ray's work of 376 pages of text, most of which consist of verbal descriptions of the fish pictured in the plates.

Nick Grindle has argued that Ray, like Lister, believed that an ordered arrangement of pictures of the external, easily visible parts of natural bodies could, alone, do the work of differentiating species.[27] But this assertion does not consider the crucial role of the textual descriptions included in Ray's *Historia Piscium*. Looking at the words and pictures together, we can see that Ray did not always see the pictures alone as sufficient for the purposes of identifying species. Consider, for example, the first plate, featuring species

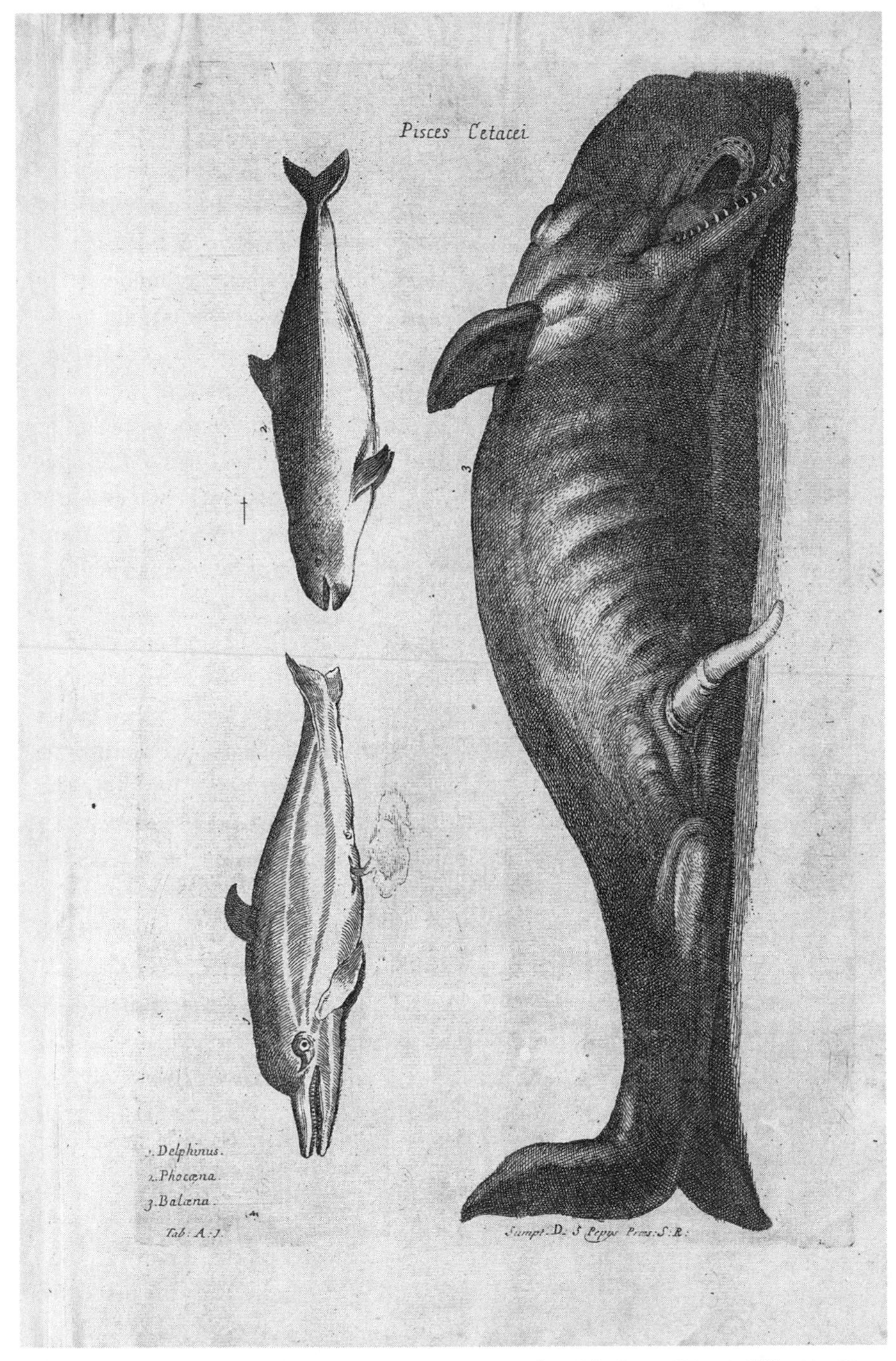

FIGURE 11. Engraved folio plate depicting cetacean species, from John Ray and Francis Willughby, *Historia Piscium* (Oxford, 1686), Tab. A 1. By kind permission of the Syndics of Cambridge University Library.

of cetaceans. It depicts, in profile, the entire external aspect of the dolphin, porpoise ("phocoena"), and whale (figure 11).[28]

For Ray, this plate was almost completely ineffective as a tool for communicating the crucial points of difference between cetaceans and fish. As he explained in the text, the close external resemblances between them concealed enormous differences. These differences revealed cetaceans to be, in key respects, closely related to what Ray called "quadrupeds," lacking the modern term *mammal*.[29] The case of aquatic mammals suggests, then, that Ray did not approve of the pictorial strategy employed by Lister and Plukenet as fully as he indicated in his review of the latter's *Phytographia*. While pictures were capable of powerfully affecting the mind, they were insufficient for telling apart natural kinds. For Ray, the most crucial distinguishing features of natural kinds could be found in a variety of internal and external features, and, as I will show, he used a concomitantly wide range of strategies—both verbal and pictorial—to represent them.

The Representation of Species and the Nature of Sensory Experience

Cetaceans provided Ray with a striking example of the difficulties that could arise from the attempt to differentiate and represent natural kinds based on a single way of seeing and representing. This may be seen by taking note of the surprising role that they played in the dispute that arose over Ray's approach to plant classification in his *Synopsis Methodica Stirpium Britannicarum* (1690). In a supplement to the second edition of that work, released in 1696, Ray reminded his critics that "sea monsters" frustrated classifications based on a simple view of the creatures in their external aspect. While an external view suggested a close kinship between them and fish, an internal view revealed them to belong to another group of species altogether.[30] It would be a mistake, however, to imagine that Ray's position came simply from the empirical observation of species such as whales and porpoises. Instead, Ray's responses to criticism of the *Synopsis* reveal that his approach to identifying and representing species was shaped not only from his own observations but also by his conception of experience itself. Along with key figures such as Boyle, Hooke, and Locke, Ray saw little formal resemblance between mental ideas and external things. His approach to representation reflected this conception of experience, seeing little necessary resemblance between words, pictures, and the things for which they stood.

The botanists Joseph Pitton de Tournefort (1656–1708) and Augustus Quirinus Rivinus (1652–1723) both found fault with Ray's approach to plant classification soon after the publication of his *Synopsis* in 1690. The two botanists thought Ray's classification was flawed because they regarded it as internally inconsistent. Sometimes Ray had grouped plants according to the similarities and differences identifiable in their reproductive parts—the fruits and flowers. But in some cases, however, he had relied on a wide range of other aspects of plant morphology to devise his groupings or attribute one plant or another to a particular class. As Phillip R. Sloan has shown, Tournefort and Rivinus believed that the only sure way to identify the species of plants was to consistently judge them by the morphology of their reproductive parts. Indeed, Tournefort argued that even where there were other signs of similarity or difference, only the forms of reproductive parts could serve as a sure guide to the identification of species. Both he and Rivinus saw the fruits and flowers of plants as "essential parts"—visible signs of the invisible essences or properties that caused plants to belong to one species or another. Ray's approach was therefore perplexing to his two critics. Sometimes he appeared to have grouped plants into species according to the differences identifiable in what Tournefort and Rivinus saw as essential parts. But at other points he appeared to have abandoned the scrutiny of essential parts in favor of whatever morphological signs had seemed most convenient.[31]

Ray responded to this criticism by releasing a second edition of his *Synopsis*, including his correspondence with Rivinus, and by dealing with Tournefort's remarks in a pamphlet entitled *De Variis Plantarum Methodis Dissertatio Brevis* (1696). In this pamphlet, Ray revealed that his conception of experience precluded the possibility of ever knowing the essential properties of any external objects. There was simply no way, he argued, for the organs of sensation and cognition to furnish the mind with ideas capable of revealing, with certainty, the essential properties of things. It was therefore futile to classify plants or any other natural kinds by systematically comparing the morphology of supposed essential parts. There was no way to know for sure that the visible similarities and differences in those parts corresponded to the invisible, essential properties of the things to which they adhered. Ray stated this position with clarity in the *Dissertatio*, right at the beginning of the section refuting the possibility of knowing the essential parts of plants:

> The essences of things are entirely unknown to us. Since all our knowledge derives its origin from sensation, we know nothing of the things which are outside us except through their ability to make impressions on our senses in some particular way, and by the mediation of these impressions to stir up some phantasm or other in the intellect.[32]

Ray asserted that all mental ideas of external things had their origins in sensation, and this fact made the essences of things entirely inaccessible to the mind. There was no possibility of unmediated knowledge of the external world. Mental ideas resulted from impressions made by external things on the senses, and the subsequent operations of the senses on the faculties of the mind. With his use of the word *phantasm* ("phantasmata"), moreover, Ray hinted that the faculty most closely involved with turning sensory experience into mental images was the imagination.

Ray immediately moved on to discuss the most important consequence of his argument for the idea of identifying species by essential parts. He began, as he had in the previous paragraph, by discussing the possibility of knowing essential properties in general, suggesting that the most likely explanation for essential qualities was the corpuscular hypothesis. As we saw in chapter 2, this position held that the perceptible forms taken by matter were caused by the interactions between undifferentiated, fundamental particles of matter known as atoms. In Boyle's case, those atoms were grouped into larger—but still imperceptible—clusters known as corpuscles. Although Ray expressed himself cautiously in the *Dissertatio* itself, he certainly concurred with this position, at least in its outlines. In a letter of 1685 to Tancred Robinson, he had written that he could not "imagine any other Difference of Bodies, but what proceeds from the Motions [or] Figures of their component Particles."[33] As Ray made clear with these words, and as we once again saw in chapter 2, the corpuscular hypothesis could never be expressed as a certainty—only as a probability—because the particles themselves were too small to be perceived by the senses. The only empirical evidence of their existence and workings was indirect, derived from inferences taken from visible phenomena such as Boyle's famous chemical experiments. For Ray, the corpuscular hypothesis thus depended on the likelihood, girded by the physico-theological assertions that we encountered in chapters 1 and 2, of an analogy between things that were amenable to sensory experience and those that were not. In the *Dissertatio*, Ray consequently explained that the inability of the senses to perceive individual "minima"—his word for the fundamental particles on which nature depended—made it impossible to truly know the essential properties of things:

> Since we cannot perceive those minima individually, no matter how the senses are augmented or aided, it is certainly necessary that their form or proportion remains concealed from our senses.[34]

In making this point, Ray was not trying to suggest that the external parts of plants were incapable of yielding clues about their internal constitution. Instead, he was simply trying to show that sensory experience posed an insuperable barrier to completely certain knowledge of the essential qualities of things. It was therefore improper to define the reproductive parts of plants as "essential parts." Even if those parts could be used to help group plants into species, no one could ever know for sure that they truly indicated essential differences.

Sloan has shown that Ray's conception of the relationship between experience and reality effectively made him a species nominalist rather than an essentialist. Even though Ray believed that natural kinds really existed, he thought that the classifications made by humans corresponded primarily to ideas generated from sensation. Those mental ideas, as we have seen, furnished only indirect evidence of the qualities inherent in external things.[35] Rather than pursuing Sloan's interest in the history of classification, I wish to highlight the consequences of this conception of experience for Ray's attitude to representation and its effects on the mind. All those consequences flowed from the insistence, so economically expressed by Ray, that there was a lack of resemblance between mental ideas and the things for which they stood.

This principle of nonresemblance found its most famous statement in John Locke's *Essay Concerning Human Understanding*. In that famous exercise in empiricism, Locke distinguished between what he called "primary qualities" and "secondary qualities." Primary qualities, he argued, were mental ideas that truly resembled external things and could not be distinguished from them. Roughly speaking, those properties corresponded to the most obvious external attributes of things—"Solidity, Extension, Figure, and Mobility." Like Ray, however, Locke contended that most of the qualities attributed to external things by the mind did not in fact resemble those things at all. Locke called those "secondary qualities." They did not inhere in things but were instead caused by the effects of imperceptible particles of matter on the organs of sensation and cognition. Colors, for example, did not belong to objects but were instead caused by some imperceptible "texture, that hath the power to produce such a sensation in us."[36] Locke took care, as Peter Anstey has shown, not to promote the corpuscular hypothesis—or any other atomistic theory of matter—in the *Essay*. It is nevertheless clear, as Anstey and others have pointed out, that the distinction between primary and secondary qualities had its origin in just such hypotheses. Locke's account closely resembles arguments made by both Boyle and René Descartes concerning the differences between our sensory perceptions and the real qualities from which those perceptions spring.[37] In turning to the *Essay*, I do not therefore mean to suggest that Locke adopted precisely the same theory of matter as Ray and his fellow naturalists. My point instead is that, as Ray's remarks in the *Dissertatio* and scattered reflections from Grew and Hooke demonstrate, those naturalists broadly agreed with Locke's account of the difference between what we perceive and the reality of things. Examining Locke's *Essay* may therefore help us understand Ray's conception of experience and its implications for his approach to the identification and representation of natural kinds.[38]

The principle of nonresemblance was central to Locke's reflections on the identification and representation of natural kinds. One such reflection can be found in a chapter on the "Remedies of the Imperfection and Abuse of Words," in book 3 of the *Essay*. Locke began by asserting that an external view of a given species—a view corresponding to Lister's pictorial approach—would be far more effective than words for conveying clear images of the shapes of things to the mind. At this point, however, he changed tack, arguing that many ideas other than those of external shapes were needed to obtain an adequate conception of the differences between one species and another. Those differences, moreover, could not easily be represented: "many of the simple *Ideas* that make up our specific *Ideas* of Substances, are [. . .] not obvious to our Senses in the Things as they ordinarily appear."[39] For Locke, the best approach to identifying the differences between species was therefore historical. It was impossible, he argued, to be sure that one kind of sensory experience gave the mind access to the essential qualities of a given natural kind. Since the shape—a primary quality—was not enough on its own to distinguish one species from another, the philosopher or naturalist had to rely instead on the enumeration of secondary qualities that had scant resemblance to the object that caused them. In formulating this position, Locke owed a good deal to Boyle. As Anstey has demonstrated, Locke formed his ideas about the value of natural history through a close engagement with Boyle's historical and experimental programs of research into the otherwise imperceptible properties of the air. Perhaps more important for our purposes, moreover, John W. Yolton has shown that Locke drew heavily on the approach to distinguishing between natural kinds laid out by Boyle in the *Origine of Formes and Qualities*.[40]

We can illustrate both Locke's argument and its relationship to Boyle's account of natural kinds by turning to the example with which Locke proved his point—gold. Locke found gold thought-provoking because it was usually identified by its color—a secondary quality. To have a clear idea of the shape of a piece of gold, by contrast, would do little to help distinguish it from other natural kinds. Locke asserted that the color of gold was best communicated through direct experience. To possess a clear idea of what distinguished that metal from so many others, however, it would be necessary to enumerate a range of additional secondary qualities that were either difficult or impossible to represent visually:

> For he that, to the yellow shining Colour of *Gold* got by sight, shall, from my enumerating them, have the *Ideas* of great Ductility, Fusibility, Fixedness, and Solubility, in *Aqua Regia* [nitro-hydrochloric acid], will have a perfecter *Idea* of *Gold*, than he can have by seeing a piece of *Gold*, and thereby imprinting in his Mind only its obvious Qualities.[41]

Locke took this list of secondary qualities, almost word for word, from Boyle's *Origine of Formes and Qualities*. He did not use it, however, to illustrate quite the same point. For Boyle, this list of accidents showed that the scholastic doctrine of forms—in which substances were defined by immaterial essences—had no basis in experience. What people really did when they described species, Boyle argued, was simply to enumerate a large enough number of perceptible accidents to reliably distinguish one substance from another. Locke agreed with this pragmatic approach to the identification of natural kinds. He made the gold example do a little more work, however, by arguing not only that the distinction of species required the enumeration of a large number of secondary qualities but also that those qualities might not be accessible to the most obvious forms of sensory experience. As Ray pointed out in the case of cetaceans, the impossibility of knowing the essential properties of things by means of sensory experience made it necessary to gather a large number of observations, both obvious and less obvious, to discriminate among species.[42]

Like Boyle and Ray, Locke approached the representation of natural kinds by focusing on the lack of resemblance that he had identified between the ideas resulting from sensation and the imperceptible arrangements of particles that had caused them. In the same place where he discussed the representation of gold, Locke reminded his readers of this point, explaining that "the whole extent of our Knowledge [. . .] reaches not beyond our own *Ideas*, limited to our ways of Perception."[43] On this basis, there could be no guarantee that it was enough to view something in its external aspect, or make a representation that attempted to reproduce that kind of sensory experience. In many cases, pictorial representations needed to be supplemented by other representations, both verbal and visual, that sought to reproduce the ideas resulting from other kinds of sensory engagement. What Locke's discussion helps us see, I think, is that Ray and many (but not all) of his fellow naturalists had powerful philosophical motivations for accompanying pictures with verbal descriptions. They did not resort to descriptions simply because their desire to use pictures was thwarted by the high cost of having engraved plates made. They were also motivated by a conception of experience that, in most instances, made the production of pictorial representations that truly resembled their objects impossible. Ray's rejection of the pictorial approach of Lister and Plukenet stemmed, I would suggest, from his belief that sensory experience could only provide indirect evidence of the real properties of things. It was therefore a fatal error, as he saw it, to depend on representations depicting a single feature or group of features of a given class of natural things. Instead, like Locke and Boyle, he argued that the best way to identify species was through the enumeration and comparison of a relatively large number of the characteristics amenable to sensation. The use of many characteristics would mitigate the possibility that some of them were misleading guides to the reality of things.[44] Alongside a picture showing the external aspect, this enumeration would result in a kind of composite representation of the species in question. Rather than resembling the species in a straightforward way, this composite representation expressed various forms of sensory engagement, from the visual experience of the external aspect to a variety of perspectival shifts, and even experimental interventions.

Ray did not, therefore, privilege pictures for their ability to resemble the things for which they stood. He pursued an approach to representation that was driven by the principle of nonresemblance between ideas and external things. Since mental ideas did not resemble external things, it was by no means always desirable to employ a strictly mimetic approach to representation. Instead, as we will see further on, the task of representation was to produce effects in the mind that reproduced or mimicked the effects resulting from the sensory experience itself. The power of pictures came not from their greater resemblance to the things that they stood for, but rather from their capacity to affect the mind in a powerful way. This was the point that Ray made in his enthusiastic remarks of 1684 about the power of picturing to Tancred Robinson. Pictures were effective because they could reproduce the suddenness, ease, and pleasure of visual experience, resulting in a vivid and lasting mental image. Yet naturalists did not think pictures alone could reproduce the pleasures of sensory experience. Rather, as Ray and Locke's reflections on the limits of pictorial representation suggest, they thought that words could sometimes be used to greater effect than pictures. Indeed, they both thought that verbal descriptions could sometimes induce similar sensations of ease, pleasure, and suddenness in the mind. In a word, they thought it *possible* to reproduce the sensuous, pleasurable dimensions of experience through verbal descriptions. The pleasures of the senses were central to their descriptive project.

Verbal Picturing and Experience

One of the descriptive strategies that Ray most often used was the comparison. Let us consider just one striking example, taken from his *Historia Piscium*. The description in question seeks to differentiate a species of fish called the thornback ray (*Raja clavata*) from its close cousins, including skates. Ray gave skates the name "Raia lævis undulata," but today they are classified as an entire family of species called Rajiformes, which also includes stingrays and guitarfishes. Ray identified what he saw as a key difference between the two kinds. Skates have teeth that are fairly similar to human teeth, while thornbacks are equipped with row upon row of flat, cartilaginous teeth adapted for crushing their prey. This difference was difficult to represent in the engravings since it was located in the inside of the mouth. To give his readers a sense of what these odd structures were like, he turned to two comparisons: "The mouth [of a thornback] is without teeth. In truth the jawbone is criss-crossed with rude rhomboidal protuberances, resembling carpenters' files."[45] With these comparisons the reader is enabled—successfully, I think—to imagine a feature of the thornback in Ray's morphology. The shape of each toothlike structure is first likened to a rhombus, leading us to imagine a structure with that figure. Next, the comparison with a carpenter's file makes us imagine the rhomboidal structures crisscrossing each other, and even helps us grasp their grinding function. Here, Ray pictured things *verbally*, projecting images in a manner that reflected his relative lack of concern for mimesis. We are led to imagine the structure of the thornback's jaw by comparison with other objects that are in many respects dissimilar to it—especially the carpenters' files.

In chapter 2, we encountered comparisons like these, seeing that Boyle sometimes regarded them as philosophically rigorous analogies, necessary for making otherwise inconceivable things conceivable. Such was the requirement of a combined physico-theology and natural history that insisted on deriving all its inferences from evidence that could be taken in by the senses. The only way to make invisible or otherwise imperceptible things amenable to sensation was through the substitution of other objects (or representations of them) that were taken to be similar enough to permit the mind to grasp an aspect of their form or function. It was for this reason that Boyle, as we saw, compared the possible mechanisms of the resurrection of the body to a chemical experiment involving camphor. These comparisons, however, were also rhetorical devices, the purpose of which was by no means limited to the strictly epistemological work of analogy. Their persuasive power came in part from their ability to stimulate recollections of easy, vivid, and pleasurable sensory experiences.

In rhetorical theory, both ancient and modern, the comparison was one of several verbal strategies recommended for vividly reproducing sensory experience. Collectively, when put to the purpose of making a description particularly vivid and persuasive, they were grouped together as a rhetorical figure known by the Greek names *hypotyposis* and *enargeia*, both of which imply an ideal of vividness. In his extremely influential rhetorical treatise *Institutio Oratoria*, the Roman rhetorician Quintilian translated those names into Latin with the word *evidentia*, which also suggested the ideal of a very vivid or obvious description.[46] But as Quintilian explained, *enargeia* referred to something more than the clear or vivid representation of actions or objects. It was, he insisted, far more powerful than that. As he put it, "[*enargeia*] is something more than clearness, since the latter merely lets itself be seen, whereas the former thrusts itself upon our notice." For Quintilian, *enargeia* could make readers or listeners feel as if the objects being described were actually present before them, as if displayed "in their living truth to the eyes of the mind."[47] The same ideal of descriptive writing continued to interest rhetorical theorists of the later seventeenth century. Consider, for example, its treatment in the rhetorical manual written by the French mathematician Bernard Lamy, *De l'art de parler* (1675). This work, often known as the "Port Royal Rhetoric" because of its author's association with the circle of Jansenist intellectuals affiliated with the Port-Royal Abbey outside Paris, was widely read in English translation as *The Art of Speaking* after its publication in 1676. In that work, Lamy referred to the figure by its near synonym *hypotyposis*, but the thrust of his discussion was the same as Quintilian's. With *hypotyposis*, he claimed, speakers and writers could give words the capacity to represent things as if they were actually present, to "form an Image [. . .] that represents the things themselves."[48] In both ancient and contemporary rhetorical theory, therefore, *enargeia* and its synonyms represented an ideal of description that saw words as capable of making things seem present, somehow reproducing the experience of visual sensation itself.[49]

As well as investing verbal description with the power to reproduce sensory experience, rhetorical theorists attributed to it the same affective power that so troubled the Royal Society in its attempts to reckon with rhetoric. Quintilian had argued that extended descriptions were ideally suited to enhancing the affective power of an oration. One of the examples that he gave concerned the possibility that an author could enlist sympathy for the victims of the sacking of a city with a lengthy description of their sufferings. The longer the description, he argued, and the more examples of suffering it contained, the more powerfully it would win readers over to the side of the unfortunate victims of the sacking.[50] In the seventeenth century, meanwhile,

Lamy had this to say about the power of what he called *hypotyposis*: "It is particularly the *Hypotyposes*, or lively Descriptions, which [. . .] raise in the Mind Floods of Passion, of which we make use, to incline the Judges as we have a mind to lead them."[51] When Lamy came to discuss *hypotyposis* in its narrowest sense—as a means of making things seem present—he still gave it an emotional valence, describing its effect as a "pleasant Enchantment [that] makes us believe we behold the things themselves." Likewise, in one of several discussions of the use of comparisons, he urged that they could be used to make otherwise difficult matters more easily understood and more vividly conceived by the mind. For this reason, he urged would-be orators to use comparisons that were "both sensible and pleasant," hinting again at their capacity to make things feel present and the possibility that this illusion would lead to a form of pleasure.[52] Lamy's discussion of *enargeia* is thus strikingly reminiscent of Ray's enthusiastic remarks of 1684 on the power of picturing. In those remarks, as we saw, Ray argued that pictures were effective because of the pleasurable, easy, and rapid way in which they led the mind to form ideas of things. He attributed to pictures—to visual experience—the same power of affective, vivid picturing that Lamy attributed to the form of verbal figuration known as *enargeia* or *hypotyposis*.

Carlo Ginzburg has shown that the affective power of *enargeia* had long been important in the writing of history. The humanists of the Renaissance understood history writing to be a fundamentally rhetorical exercise in moral or political exhortation, to be accomplished through the presentation of striking examples of good and bad conduct.[53] Modern descriptions of historical events are usually seen as purportedly accurate accounts of things that have taken place, backed by an accumulation of citations from reliable sources. For ancient and humanist historians, however, the idea of *evidence*—as we have seen, a synonym for *enargeia* in Quintilian's treatment—referred to the vivid representation of examples of good and bad conduct to be either emulated or rejected. Humanist historians therefore valued forms of verbal figuration that could leave readers with vivid, lifelike mental images of past occurrences. As Ginzburg reminds us, the Greek historian Plutarch wrote in his essay *On the Fame of the Athenians* that "the most effective historian is he who, by a vivid representation of emotions and characters, makes his representation like a painting."[54] The ideal of *enargeia* was thus central to the work of historical representation. With vivid descriptions, historians hoped to give readers clear mental images of the events that they described, often stripped down to a core of moral meaning rather than a strictly veracious account of things as they happened. It was through such images that they hoped to fulfil the ethical work then associated with the reading and writing of history.

It is well known, moreover, that the same poetic and rhetorical ideals were central to early modern art theory. Seeking to show that painting, despite its association with manual labor, deserved a place alongside the liberal studies, art theorists of the sixteenth and seventeenth centuries frequently modeled visual representation on ancient prescriptions for verbal expression, especially the *Poetics* of Aristotle (384–322 BCE) and the *Ars Poetica* of Horace (65–8 BCE). The result, as the art historian Rensselaer Lee showed in the early twentieth century, was that art theorists tended to identify both the aesthetic and the moral aims of painting closely with those of poetry and history. In so doing, they found authorization both in Horace's famous dictum *ut pictura poesis* (as in painting, so in poetry), along with the claim of the poet Simonides (556–468 BCE), cited by Plutarch in *On the Fame of the Athenians*, that "painting [is] inarticulate poetry and poetry inarticulate painting."[55] Of course, artists were sensitive to the differences between the two media. Leonardo da Vinci argued, for instance, that painting was superior to poetry because it could picture in an instant what poetry could only accomplish over an extended period. Nevertheless, it is fair to say that most art theorists tended to overlook that seemingly obvious distinction, asserting instead that painting and poetry produced images of a similar kind, attributing to static pictures a capacity for narrative exposition and to sequences of words a capacity for static picturing.[56] Indeed, it was not until the middle of the eighteenth century that philosophers and critics turned decisively against the notion that poetry and painting pictured things in more or less the same fashion. Most famously, the German philosopher Gotthold Lessing (1729–1781) attacked descriptive poetry in his *Laocoön: An Essay on the Limits of Painting and Poetry* (1766). Like Leonardo, he argued that words could never stand in for visual experience because vision was basically instantaneous, while words could only ever be experienced in sequence, over time. For Lessing, words were far better suited to the narration of events that, like words themselves, had unfolded over time.[57]

In the late seventeenth and early eighteenth centuries, by contrast, the ideal of picturing summed up by *enargeia* still exercised considerable influence.[58] During the 1690s, for example, the Society of the Virtuosi of St. Luke, an art appreciation club founded that decade in London, commissioned the poet John Dryden (1631–1700) to translate a treatise on painting entitled *De Arte Graphica* (1668) by the French painter Charles Alphonse du Fresnoy (1611–1668). To this translation, published in 1695, Dryden added his own "Parallel betwixt Painting and Poetry." Unlike Lessing, Dryden had little to say about the role of time in differentiating narrative from static picturing. Instead, he argued that words and images were both equally capable of accomplishing what he—like Aristotle—regarded as the fundamental aim of artistic representation:

> For *both* these *Arts* [. . .] present us with Images more perfect than the Life in any individual: and we have the pleasure to see all the scatter'd Beauties of Nature united by a *happy Chymistry*, without its deformities or faults.[59]

As Lee has pointed out, the essayist Joseph Addison (1672–1719) took a more complex position in his famous series of essays on the "Pleasures of the Imagination," published over several issues of the *Spectator* in 1712. There, Addison sometimes compared verbal description unfavorably to visual picturing, remarking both that vision appeared to be the cause of greater pleasure and that visual representations had a far closer resemblance to their objects. Despite those intimations, he nevertheless went on, much like Ray, to suggest that well-chosen words could provoke images even more vivid than those resulting from direct visual encounters. Toward the end of his final essay on the imagination, exactly reproducing the language of *enargeia*, he explained that "Words, when well chosen, have so great a Force in them, that a Description often gives us more lively Ideas than the Sight of Things themselves."[60]

Poetry and rhetoric thus infused seventeenth-century literate culture with strategies for verbally reproducing the effects of sensory experience—both the mental images that experience could lead to and the affective transformations that could accompany them. At the same time, and perhaps more important, those strategies came to inform discourses about what experience itself could be like—as the example of art theory helps show. It would be tempting at this point to attribute the slippage between words and pictures in the work of Locke and Ray to the pervasive influence of rhetorical culture in the seventeenth century. This influence does furnish a kind of general explanation. Ray and almost all his contemporaries in the Royal Society had received a rhetorical education and were keenly aware of the strategies of rhetorical persuasion. It is therefore hardly surprising that Ray's comments on the power of words and pictures conform so closely to rhetorical archetypes. But the society's own insistent rejections of pleasurable rhetoric serve as a rebuke to any attempt to infer straightforward connections between the prevailing rhetorical culture and the forms of rhetorical expression used by naturalists of the seventeenth and early eighteenth centuries. Ray and his colleagues did not deploy figures such as the comparison simply because of their familiarity with the rhetorical ideals of representation that were then so influential. They used comparisons because the ideal of *enargeia* both reflected and informed their ideas about some of the processes by which they believed the mind could form clear images of things.

The Pleasures of Comparison

An example of the place of *enargeia* in the theories of mind and models of cognition used by Ray and his contemporaries is to be found in Locke's *Essay*. In a chapter delineating several of the operations of the mind, Locke set out to argue that individuals needed to exercise "judgment"—the mind's capacity for distinguishing between the ideas generated by sensation—if they were to discover the truth. Without judgment, he argued, the mind would be incapable of distinguishing truth from falsehood amid the mass of sensory impressions by which it was continually bombarded.[61] To drive home his point, Locke contrasted judgment with another mental operation—"wit." Unlike judgment, wit consisted in the identification of similarities between ideas. And rather than taking reason as its guiding principle, wit was governed by pleasure. Locke therefore contrasted the pleasures of wit with the hard work of judgment, cautioning his readers that only the latter could lead to useful inferences about the external world:

> For *Wit* lying most in the assemblage of *Ideas*, and putting those together with quickness and variety, wherein can be found any resemblance or congruity, thereby to make up pleasant Pictures, and agreeable Visions in the Fancy: *Judgment*, on the contrary, lies quite on the other side, in separating carefully, one from another, *Ideas*, wherein can be found the least difference, thereby to avoid being misled by Similitude, and by affinity to take one thing for another.[62]

The problem with wit was that the enjoyable comparisons generated by it were of little value to the pursuit of truth. The demands of pleasure, the animating principle of wit, could be satisfied by likenesses that were merely apparent or superficial, concealing other differences that might be harder to identify. Only by painstakingly searching for the differences between ideas could one be sure of working out which of them belonged together. This point echoes the practical example of Ray's interest in the classification of cetaceans. Cetaceans and fish resembled each other in obvious ways that were easy to apprehend, but a deeper, more difficult inquiry revealed that they had few real similarities.[63]

While Locke asserted the superiority of judgment over wit, he nevertheless intimated that wit's "pleasant Pictures" could exercise great power over the mind.[64] He immediately confirmed this impression by arguing that metaphor, which he simply characterized as the operation of verbally comparing one thing to another for pleasurable purposes, depended on the operations of wit. In so doing he effectively produced a general explanation for the affective power of rhetoric:

> [Judgment] is a way of proceeding quite contrary to Metaphor and Allusion, wherein [. . .] lies that entertainment and pleasantry of Wit, which strikes so lively on the Fancy, and therefore so acceptable to all People; because its Beauty appears at first sight, and there is required no labour of thought, to examine what Truth or Reason there is in it.[65]

The operations of wit gave rhetoric its affective power. The imagination, according to Locke, took pleasure in the kinds of quick, easy, and pleasurable comparisons facilitated by metaphorical speech. And it was by these operations that the imagination formed images that were so vivid and powerful as to overpower the scruples of reason.[66] Locke's account of metaphor reproduced the most important elements of the descriptive ideal of *enargeia*. He suggested that words could be used to reproduce visual experience and that the resulting images possessed a vividness that was either caused by or led to their affective power. At the same time, however, he departed from the descriptions of *enargeia* to be found in rhetorical treatises, both by specifying the exact cognitive process on which it seemed to depend and by insisting that this process was inimical to the exercise of judgment and discovery of the truth.

Locke's explanation for the power of metaphor helps us to grasp how important the provocation of affective states was to the descriptive strategies employed by Ray and his fellow naturalists. When they used comparisons to represent natural things, they took advantage of the mental operations that Locke saw as belonging to wit. That Ray considered this mental operation to be epistemologically useful is clear from his account of the power of pictorial representation—the one that he sent to Robinson in 1684. Ray's interest in using pictures to effect the rapid and pleasurable formation of vivid ideas closely resembles Locke's description of the "agreeable Visions" and pleasurable feelings arising from metaphor. A difficulty arises, however, when we try to square this interest in the power of wit with the fact that both men complained that the pleasures of metaphor were fundamentally deceptive. Once again, however, we find that disavowals of rhetoric concealed a deep interest in harnessing those pleasures for more worthy purposes. Peter Walmsley has shown that Locke's dismissal of wit only barely concealed his belief that it was one of the fundamental features of cognition, characteristic of the nonrational functions of the faculties of imagination and memory. Indeed, the

Essay devotes many pages to the discussion of cases in which experiences of pleasure and pain seemed to have led the imagination to form associations between ideas that could never have resulted from the exercise of judgment.[67]

It was this nonrational principle, governed by pleasure, to which the naturalists appealed when they used rhetorical devices such as the comparison. In the next chapter, examining the style of these comparisons, we will bring to light the cognitive and affective functions that were attributed to them in the interconnected discourses of neurology and rhetoric. A short reflection on a few of the comparisons reveals, however, their appeal to the pleasurable operations of wit. When Ray compared the interior of the jaw of the thornback to a carpenter's file, he was surely not appealing to the faculty of judgment. He cannot have been inviting the reader to laboriously identify the numerous differences between the sensory impressions left behind by those two things. He instead aimed to make the reader quickly recall the memory of the idea resulting from a previous sensory experience—the sight of a carpenter's file—and relate it to the thornback's mouth. Ray made use of the clarity and pleasure that could result from this operation to form an association linking the thing he was describing to an easily recalled and vivid image. The difficult form of the thornback ray's mouth would be resolved into the familiar memory of the carpenter's file, a process facilitated by the principles of ease and pleasure at work in the mental operations of wit. The same was true, to take another example, when Grew compared the parenchyma, or ground tissue of plant leaves, to "Gentlewomen's Needle-works" in his *Anatomy of Plants*, or when Hooke compared the spots covering the surface of seaweed to "the shape of the sole of a round toed shoe" in his *Micrographia*. Even Locke used comparisons in the same fashion—for example, when he compared the way in which memories gradually decay to the slow erosion of inscriptions engraved into marble.[68]

These examples suggest that the descriptive ideal of *enargeia* played an important role in the communication of natural-historical and natural-philosophical knowledge. When Ray and his fellow naturalists used comparisons to describe things, their aim was to paint vivid images in the minds of their readers. This picturing was premised neither on pictorial representation nor on a strictly mimetic understanding of the resemblances between ideas and their objects. The comparisons instead worked by provoking the mind into recalling the memory of another sensory experience—one sufficiently like that generated by the experience of the thing itself to stand in for it. The correspondence between the combination of impressions that they could leave in the organs of sensation and cognition was therefore more important than the strict resemblance between the two objects themselves. Crucially,

Locke argued that pleasure was crucial to the success of this operation. It was the pleasure taken by the mind in the easy identification of similarities between things that made the images resulting from metaphor so vivid and persuasive. The plainness of the descriptive style did not, therefore, guarantee the dispassionate communication of knowledge. Instead the evocation of pleasure, elicited through forms of figuration such as *enargeia*, was crucial to the empirical project.

This emphasis on pleasure throws the usual account of the Royal Society's attempt to tame the passions in the interest of objectivity into doubt. The successful communication of knowledge depended not simply on disciplining experience but rather on the cultivation of a particular kind of subjectivity—a feeling of pleasure accompanying the easy formation of vivid mental images. In Locke's analysis, after all, the explanatory power and persuasive force of metaphors were enhanced by the fact that they caused pleasure. But how could the objectivity be guaranteed if it depended on the provocation of a feeling so subjective as pleasure? In asking this question I do not mean to deny that Ray and his contemporaries wanted their representations to lead, in some sense, to universally valid, objective knowledge. Yet they were, as I will show in the next chapter, deeply troubled by the possibility that some of their readers might not wish to experience, or be capable of experiencing, the pleasures so necessary for the acquisition of knowledge. It was for this reason, I suggest, that they sought to guarantee the objectivity of their knowledge by appealing to, and attempting to shape, the affective dispositions of their readers. They saw the capacity to take pleasure in the study of natural history as a kind of bodily symptom of the moral and intellectual dispositions necessary for the virtuous study of nature. Ray and his fellow naturalists used comparisons and other vivid forms of representation, I suggest, to enable their readers to experience natural history and anatomy in the way that they deemed appropriate. Their verbal descriptions were tools for producing the subjectivity requisite for the acquisition of knowledge.

5

Natural Philosophy and the Cultivation of Taste

In the first paragraph of his first published work, *Catalogus Plantarum circa Cantabrigiam Nascentium* (Cambridge, 1660), a catalog of the plants growing around Cambridge, John Ray described the pleasures of botany: "We became [. . .] delighted by the shape, the colour and [. . .] wonderfully graceful outward appearance of individual plants."[1] For Ray, it was perhaps obvious that others would share his delight in studying the bodies of plants and animals. He regarded them, after all, as the material manifestations of God's wise and beneficent care for the world. Sometimes, therefore, Ray asserted that the encounter with design in nature was bound to cause pleasure. In *The Wisdom of God Manifested in the World of Creation*, he almost described that pleasure as a reflex, readily observable in human responses to the beauty on display in the works of creation. In a passage dealing with the presence of geometrical figures in natural bodies, for example, he simply remarked that "if nature shape any thing but near to this *geometrical* accuracy," then "we take notice of it with much content and pleasure."[2] In passages such as this, there was no room for dispute. The sensory pleasure attendant upon the encounter with design in nature was a fact, witnessed by universal testimony.

Such attempts to link the production of knowledge to feelings of aesthetic pleasure were frustrated, however, when people did not experience nature in the way that Ray and his contemporaries hoped. Indeed, the weakness of Ray's assertions is revealed by the difficulties he encountered when trying to deal with the simple fact that some people did not seem capable of enjoying the study of nature as much as they should. In such cases, he sometimes resorted to an uncharacteristic rudeness. Once again borrowing his words from Henry More, Ray at one point characterized an individual incapable of delighting in nature's design as being "sunk into so forlorn a pitch of degeneracy that he is as stupid to these things as the basest of beasts."[3] For the most part, however, he and his contemporaries took a gentler tone, seeking to encourage rather than chastise. In his descriptions of chemical experiments, as we have seen, Boyle often remarked on the "delight" or "pleasure" that he had taken in the phenomena on display, whether the production of a beautiful color or the surprising transformation of one substance into another. When the pleasures in question were by no means obvious, such encouragements sometimes look a little ridiculous. In one incongruous passage, Ray unconvincingly maintained that it was possible to enjoy the spectacle of peristalsis in the esophagus of a cow: "The mention of the peristaltick Motion puts me in mind of an ocular Demonstration of it in the Gullet of Kine when they chew the Cud, which I have often beheld with pleasure."[4]

As we saw at the start of this book, the conventional narrative holds that the aesthetic claims of physico-theology served mainly in an apologetic role, having little to do with the work of natural history and natural philosophy themselves. It should already be clear, however, that Ray and his contemporaries were much more interested in harnessing the pleasures of the senses and imagination than they sometimes admitted. In chapter 3, we found that Nehemiah Grew and Robert Hooke allowed themselves to be guided by their aesthetic expectations in the investigation of things that seemed too ugly or imperfect to have been designed by God. In chapter 4, meanwhile, we saw that one of their most important representational strategies—the comparison—was widely thought to depend on the pleasures of the imagination. The fact that Ray and his contemporaries sometimes premised the production of knowledge on sensory pleasure, however, is obviously difficult to square with their evident commitment to what we now call objectivity. Indeed, their interest in mobilizing pleasure for epistemic purposes sits uneasily alongside the usual argument that the Royal Society sought above all to neutralize the distortions arising from the subjective character of much individual experience. To explain how natural philosophers could put the pleasures of sensory and imaginative experience in the service of natural philosophy, we must therefore find another way of talking about how they hoped to obtain consensus about the meanings of experience.

In this chapter, I will suggest that the vocabulary of aesthetic judgment enables us to grasp how Ray and his contemporaries reconciled the desire for objectivity with their recognition that the affective dimensions of sensory experience were central to the work of knowledge production. In making this claim, I do not mean to suggest that their ideas about the relationship between beauty and sensory pleasure were identical to those that would emerge during the eighteenth century—for instance, in the works of Immanuel Kant. The central aesthetic claim of physico-theology, as we have seen, was that that the rational discovery of nature's purposes would lead to an innocent form of sensory or imaginative pleasure. Late in the eighteenth century, by contrast, Kant argued in the *Critique of the Power of Judgment* that physico-theology was wholly incompatible with aesthetic judgment, arguing that the attempt to discover purposiveness in natural things would make the experience of beauty impossible. As Kant saw it, the rational work of discovering purposiveness in nature would deprive the mind of the freedom that it needed to exercise the faculty of aesthetic judgment. In some respects, therefore, Kant's *Critique* served as an explicit rebuke to the aesthetic claims associated with physico-theology.[5]

The emergence of aesthetic theory in the eighteenth century brought with it, however, useful vocabularies for describing the predicament that Ray encountered in *The Wisdom of God*—the desire to claim a kind of universality for subjective forms of experience. In the *Critique of the Power of Judgment*, for instance, Kant argued that the experience of beauty depended on the free play of the imagination and understanding. The formation of aesthetic judgments was therefore a subjective process, incompatible with any exercise that involved the prior determination of a concept, such as the identification of a need or the exercise of reason, that would constrain the freedom of the mental faculties. Yet Kant also claimed that aesthetic judgments had a kind of necessity about them. He suggested that a person who formed a judgment of beauty in the right manner could expect that judgment to have universal validity, even if others were not necessarily capable of sharing it. As we saw in the introduction to this book, Kant therefore argued that beauty had "subjective universality," characterizing aesthetic judgment as a subjective process that could nevertheless result in the sort of universality to be expected from the exercise of reason.[6] For the physico-theologians, as I have already suggested, the experience of beauty was quite different from what Kant would later describe, allied far more closely to the exercise of reason and—paradoxically—to the bodily pleasures of the senses and imagination. Nevertheless, we shall see that naturalists such as Ray and Robert Boyle hit upon similar experiential dilemmas when claiming that the encounter with design in nature should lead to feelings of pleasure. They needed to square their belief that the pleasures of studying nature were universal with the fact that some people did not appear capable of enjoying them. Ray and his contemporaries wanted, in other words, to show their readers how to improve their taste—that is, to take pleasure in what they should already have been able to enjoy.

This chapter focuses on three interconnected themes. The first concerns the resources available to naturalists for explaining why pleasure varied from person to person, along with the strategies on which they could call for overcoming such differences. Investigating works such as Boyle's *Some Considerations Touching the Style of the H. Scriptures* (1661), we will see that rhetorical theory gave naturalists a language for describing both intersubjective difference and the means for accomplishing intersubjective agreement. Building on the connections identified in chapter 4, the second theme concerns the links between rhetoric and neurophysiology, examining how naturalists and rhetorical theorists alike mapped the pleasures and pains of rhetoric onto the physiology of the brain and nervous system. Drawing on Bernard Lamy's *Art of Speaking* and turning once again to Thomas Willis's *Cerebri Anatome*, we will see that naturalists construed what we would call the cultivation of taste as an embodied process, linking the development of the capacity to experience appropriate forms of pleasure to the workings of the bodily organs of sensation and cognition. Finally, drawing these threads together, this chapter completes our reconsideration of what it meant to use the organs of sensation and imagination as instruments of knowledge production. Reckoning with the neurophysiological effects that naturalists sought to provoke through their descriptions of plant and animal bodies, we will see that the plain style was concerned with provoking the bodily pleasures associated with imaginative experience. The purpose of this chapter, therefore, will be to show that the aesthetic claims of physico-theology were closely connected to the communication and consumption of knowledge. Boyle, Grew, Hooke, Ray, and Willis used the provocation of certain kinds of pleasure, I would suggest, to give their readers clearer ideas of things and thereby to help them cultivate a taste for the pleasures of natural philosophy.

The Wisdom of God and the Cultivation of Taste

The threat posed to the physico-theological project by the possession of inappropriate affective states is exemplified by a debate that took place in the decades immediately before and after 1700 about the appearance and purposes of the landscape. That debate, which we briefly encountered in chapter 3, was provoked by the philosopher and clergyman Thomas Burnet, who published the two volumes of the English version of his *Sacred Theory of the Earth* in 1684 and 1690. In that work, Burnet attempted to assign physical causes to the various mutations the earth had undergone in the history laid out in the Bible, such as the Deluge survived only by Noah, his family, and the selected animals.[7] Consistent with the approach to divine providence adopted by Nehemiah Grew and other followers of Isaac Newton, Burnet also claimed that those causes, and their resultant manifestations in the form of the earth at various times, were the result of a providential plan engineered by God at the beginning of time to reward humanity for its virtues and punish it for its sins. So, for Burnet, the form of the earth—and its mountains in particular—exhibited signs of humanity's sinful estate, possessing neither use nor beauty.[8] Even as Burnet used the *Sacred Theory* to advance an attempted demonstration of God's activity in nature, he was perhaps unwittingly undermining the premises of physico-theology. As well as arguing that mountains were utterly useless to the preservation of life on earth, he asserted that they lacked entirely in beauty, calling them "wild, vast and indigested heaps of Stones and Earth." And even as he asserted that mountains were "pleasing to behold," he associated them with an affective state that was totally at odds with what the proponents of physico-theology saw as appropriate. Instead of the temperate and gentle pleasures described by Ray in *The Wisdom of God*, Burnet spoke of the "great thoughts and passions" inspired by the immense size and ruggedness of mountains.[9]

The debate over Burnet's *Sacred Theory* is still best known from Marjorie H. Nicholson's influential study of the emergence of the aesthetics of the sublime in English literature, *Mountain Gloom and Mountain Glory: The Development of the Aesthetics of the Infinite* (1959). There, Nicholson showed that Burnet's descriptions marked a crucial turning point in literary history. Burnet's evocations of the horror mixed with pleasure induced by the sight of mountains, Nicholson argued, both anticipated and influenced the emergence of sublime imagery more familiar to us from the poetry of romantics such as Samuel Taylor Coleridge, with its emphasis on nature's grandeur and power.[10] Many of Burnet's contemporaries, however, took issue with his suggestion that the landscape was somehow less beautiful than any other of God's creation. Their responses to Burnet's hypothesis can help us see how natural philosophers on both sides of the debate linked the possession of an appropriate affective disposition to the discovery of the truth about purposiveness in nature.

When not focusing on Burnet's model of physical causation or his troubling readings of scripture, physico-theologians such as Ray responded by attempting to counter his assertion that mountains were devoid of uses. In his *Three Physico-Theological Discourses* (1693), Ray thus listed the many different purposes that they appeared to serve, noting for instance their role in the water cycle and how they apparently provided habitats for a wide variety of animals.[11] But they also dismantled Burnet's aesthetic and experiential claims, countering them with assertions of the beauty of mountains and the sensory gratification that they could bring forth. As Ray saw it, the landscape was "a very beautiful and pleasant object, and with all that variety of Hills, and Valleys, [. . .] far more grateful to behold, than a perfectly level Country without any rising [. . .] to terminate the sight."[12] William Derham and Richard Bentley, meanwhile, made more explicit criticisms of Burnet's capacity to judge and experience the landscape. They claimed that he had been in the wrong state to experience the beauty of mountains and that he had chosen the wrong vantage point from which to visualize their appearance. Derham, taking advantage of Burnet's relation that he had first come to see mountains as ruins during an arduous journey over the Alps, remarked that "the Hills and Vales, though to a peevish weary Traveller, they may seem incommodious and troublesome, yet are a noble Work of the great Creator."[13] For the defenders of physico-theology, the affective register of Burnet's descriptions stood in the way of the delight that ought to accompany the acquisition of knowledge about the uses of the landscape.

In the works of Robert Boyle, we can identify an exercise in the formation of taste that closely parallels such concerns about the affective states arising from the encounter with natural things thought to result from God's agency. Composed in the 1650s and published in 1661, this work was Boyle's *Some Considerations Touching the Style of the H. Scriptures*.[14] In this book, Boyle's aim was to demonstrate that the Bible possessed a fine style, making it pleasurable to read. But the *Style of the Scriptures*, like *The Wisdom of God*, is haunted by the possibility that readers might not be able to feel the pleasures that (as seemed obvious to Boyle) should accompany the experience of reading or hearing the scriptural texts. The similarities between the two works do not end there. The tasks that Boyle set himself in the *Style of the Scriptures* were thus very much like those that preoccupied Ray. With a mixture of rational argumentation and affective exhortation, he aimed to show that the Bible's style induced pleasure. He supplied, moreover, advice about how to experience that pleasure, evoking the ideal affective state with a lengthy rhapsody toward the end of the work. The *Style of the Scriptures* thus functions as an exercise in the cultivation of taste for the Bible.

The *Style of the Scriptures* consequently contains many acknowledgments of the possibility that the Bible's style might seem unpleasant to some readers. For about the first half of the book, indeed, Boyle listed some of the many reasons that might lead readers to find the scriptural style unappealing. Rather than admit that the sacred style was actually unpleasant, however, he sought to explain its seeming deficiencies by invoking the rhetorical strategy known as *accommodation*. Accommodation was a standard part of early modern rhetorical training, and its aim was to show speakers and authors how to adjust the style of a discourse to the needs of its likely audience. As John T. Harwood has pointed out, Boyle recalled having learned about accommodation when receiving his education in Geneva during the 1650s, describing it as a matter of adjusting the style of a discourse to "the various circumstances considerable in the matter, the speaker, and the hearers."[15] To practice rhetorical accommodation, then, was to acknowledge that the merits of a given style were relative to the situation. The success or failure of a discourse depended on the extent to which the speaker successfully adjusted her or his style to the taste of the audience and the requirements of the subject in question.[16]

Using this aspect of rhetorical theory, Boyle could argue that apparent failures in the style of the Bible were in fact the result of its highly successful accommodation to the tastes of people living in different times and places. To take just one example, Boyle suggested that the "Figurative, and (oftentimes) Abrupt way of Arguing" found in some parts of the Bible was one of the stylistic qualities that had made it effective in proselytizing the people of the ancient Middle East. For Boyle, however, it was the same quality that made those parts of the text ring false in the ears of seventeenth-century gentlemen, whose customs and tastes were quite different.[17] Elsewhere, Boyle reminded his readers that the works of fashionable Roman poets such as Juvenal and Martial were only intelligible to educated readers of the seventeenth century because they knew enough about Roman language, history, and culture to make sense of them.[18] Boyle could therefore use the discourse of accommodation to argue that the style of the Bible had in fact been very well written. While this procedure vindicated God's talent as a writer, it nevertheless worked only by acknowledging that the readers and listeners of Boyle's own day could legitimately find some parts of the Bible unpleasant.

Boyle did not stick, however, to this suggestion that the style of the scriptures was only relatively pleasurable. Instead, he argued for much of the rest of the work that these pleasures were absolute and that everyone could be expected to enjoy the scriptural style. He therefore positioned the pleasure to be taken in the scriptures as what Kant would later define as a "subjective universal"—that is, a subjective experience that could nevertheless be expected from everyone. Boyle made this point with an unhelpful double negative: "it is very farr from being consonant to Experience, that the Style of the Scripture does make it Unoperative upon the Generality of its Readers, if they be not Faultily indispos'd to receive Impressions from it."[19] Here, he left the relativism of accommodation far behind. Instead, he suggested that failures to rightly experience the scriptural style were caused by deficiencies in individual readers. As in the argument based on the theory of accommodation, the reader was still very much a participant in determining the success or failure of a certain style. But instead of making it the writer's responsibility to adapt the style to the needs of readers, Boyle this time argued that readers needed to make themselves capable of enjoying it. He set out, in other words, to show his readers how to enjoy the wisdom of God.

In pursuit of this aim, Boyle mobilized the metaphor of taste in a highly literal manner. In a passage extolling the virtues of regular Bible reading, for instance, he used a striking medical metaphor to account for how it might be possible to lose or gain the capacity to enjoy the scriptural style. Regular reading, he suggested was a form of medical regimen that would eventually cure the distemper that stood in the way of enjoying the style of the Bible:

> 'tis not onely a Commendable, but a much more Improving Custome than 'tis by many thought, to Read daily and orderly some set Portion or Chapters of the Bible: and not to desist from that Practice, though [. . .] we should not perceive a sudden and sensible Benefit accruing from it. For in Diseases (Bodily or Spiritual) though the Mouth be out of Tast, and cannot Rellish what is taken in, yet wholesome Aliments must be eaten, and do effectively Nourish and strengthen, though they be then Insipid, (perhaps Bitter) to the distemper'd Palate.[20]

Boyle described the Bible itself as an agent that could transform the subjective experiences of patient readers, gradually giving them the capacity to enjoy it. Later, he would repeat this suggestion by characterizing the style of the Bible as an inscriptive tool that (like the writing implement from which the word *style* takes its name) could make deep impressions on the minds of readers, changing them for the better.[21] At one point, for example, he employed a passage from the Epistle to the Hebrews to convey the transformative power of the style of the scriptural style: "For the Word of God, [. . .] is Quick and Powerfull, and Sharper than any Two-edged Sword, piercing even to the Dividing asunder of Soul and Spirit."[22] The beautiful hyperbole of this latter passage masks its essential conformity with contemporary rhetorical theory. Naturalists and orators alike believed that a well-chosen style could help readers make themselves better by giving them a taste for morally improving texts that might otherwise have seemed unpleasant.

Boyle's approach to the scriptural style gives us some useful tools for thinking about the aesthetic claims of physico-theology. As we have seen, Ray also acknowledged that the pleasures of natural philosophy were subjective, noting that some people seemed either unable or unwilling to experience them as he thought proper. Much like Boyle, moreover, he suggested that the encounter with design in nature could itself serve as a cure for the inability to experience its pleasures. Toward the start of *The Wisdom of God*, Ray cautioned his readers not to worry too much if they did not yet enjoy the study of nature:

> I know that a new Study at first, seems very Vast, Intricate and Difficult; but after a little Resolution and Progress, after a man becomes a little acquainted, as I may so say, with it, his Understanding is wonderfully cleared up and enlarged, the difficulties vanish, and *the thing grows easie and familiar*. And for our Encouragement in this Study, observe what the Psalmist saith, Psal. 111. 2. The works of the Lord are great, sought out of all them that have pleasure therein.[23]

Like Boyle, Ray characterized the encounter with the wisdom of God as an exercise in the cultivation of taste, arguing that feelings of ease and familiarity would eventually replace early unpleasantness. The process of aesthetic cultivation that Ray described, however, was far more important to the work of natural philosophy than its role in encouraging the study of nature would suggest. This is a point that becomes clear if we look beyond the function of aesthetic claims in the rhetoric of physico-theology and start to look instead at the stylistic strategies at work in natural history and anatomy. Focusing on the style of Willis's *Cerebri Anatome*, we will see that the work of representing nature was also an exercise in the cultivation of taste. When Ray and his contemporaries described the bodies of plants and animals, they sought to induce forms of pleasure would in turn lead to both cognitive and moral benefits.

The Style of Willis's *Cerebri Anatome*

Until recently, as we have already seen, scholars paid little attention to the possibility that the plain style of natural-historical and natural-philosophical texts from the later seventeenth century may have been intended to provoke affective responses. Historians and literary critics instead tended to take the Royal Society's antirhetorical postures at face value, regarding the plain style a sort of antistyle, with no greater purpose than accurately transcribing the results of experience into verbal form.[24] But there is no good reason why the apparently plain style of the verbal representations made by naturalists such as Grew, Ray, and Willis should deter an attempt figure out its intended affective meaning. This point might already be clear from chapter 4, where we saw that Ray and his correspondents used strategies of verbal picturing to provoke the mind into recalling images that had been powerfully registered in the memory. If we might then wonder just how plain Ray's natural-historical style was, it is nevertheless more important here to recognize, as Jan Golinski and Peter Walmsley each have shown, that the plain style was a style that worked in the same way as any other—by attempting to resolve the competing needs of subject, speaker, and audience.[25] As Walmsley puts it, "for a generation educated on Quintilian, the plain style was a *style*—not a claim to transparent meaning or an endorsement of some linguistic purity, but a literary mode with its own appropriate purposes and strategies."[26] Even if the style of such works as Ray's *Historia Piscium* or Willis's *Cerebri Anatome* was comparatively plain, this does not mean that their authors lacked the intention to promote certain affective states or transformations.

The key to understanding the plain descriptive style, I would suggest, involves recognizing that its effects—like those of all styles—can only be grasped by thinking about how it was intended to reconcile the competing demands of subject, speaker, and audience. Bernard Lamy made this clear in *The Art of Speaking*, the rhetorical treatise in which he explained the well-known powers of rhetoric using the framework of René Descartes's physiological account of sensation and cognition. Notwithstanding this novelty, he retained the classical tripartite division of styles into "plain," "middle," and "lofty." For Lamy, however, the effects of those styles were not fixed. Instead, their effects depended in large measure on how speakers and writers used them either to raise or to lower the affective register associated with the subject at hand. Using the works of the Roman poet Virgil (70–19 BCE) as a template, Lamy gave examples of the main rhetorical styles, matching them to their subjects. The plain or low style, in which comparisons and other metaphors were simple and largely devoid of poetic amplification, was appropriate to pastoral subjects, as in the *Eclogues*. The lofty style, meanwhile, was appropriate to heroic actions such as those in the *Aeneid*, requiring elaborate metaphors to convey the grandeur of its subjects. The middle style, exemplified by the *Georgics*, was suitable to works seeking to educate or instruct readers in a pleasant fashion. It thus called for comparisons that amplified their subject enough to make them interesting or diminished them enough to make them intelligible.[27] To achieve what was known as *decorum*, the appropriateness of a style to its subject, the writer had to consider the fitness of metaphors to the things that they stood for, along with their likely effects on readers and listeners. To take another example, Joseph Addison explained in his essay on the *Georgics* (included with John Dryden's 1697 edition of the works of Virgil) that Virgil used relatively florid comparisons to describe agricultural matters because of the need to make the otherwise boring precepts of husbandry palatable to genteel readers. Even though the ostensible subject at hand—husbandry—generally called for a low style, the needs of the intended genteel audience in this case justified an elevation in the stylistic pitch.[28] Orators also needed to ensure that their comparisons and other metaphors did not stretch credulity, either by being too dissimilar to their objects or by being too incoherent with the overall temper of a work. There was no place for lofty metaphors, for example, in a work written in the low style.[29]

Lamy and Addison understood that the effects of a style should not be weighed up against fixed criteria. They could only be worked out by thinking about the relationships among styles, their subjects, and their likely effects on audiences. To work out what Grew, Hooke, Ray, and Willis expected to accomplish with their verbal representations of natural bodies, we therefore need first to find out the prevailing tone of the various forms of figuration and metaphor that they employed. Having worked out what that style was, however, we should not view it in isolation. Rather, we need to clarify its intended effects by considering its relationship to the subject and at hand and to its likely audience. To accomplish these two ends, I propose to offer a systematic analysis of the style of Thomas Willis's *Cerebri Anatome*, focusing especially on its use of comparisons to represent the fabric of the brain. There are two powerful motives for this focus on the comparison, understood not in general terms but rather as a figure or trope of rhetoric. First, the comparison is a basic metaphor that occurs frequently in works of natural philosophy, natural history, and anatomy, making it possible to compare many instances of the same form of figuration with one another and therefore to get a good sense of the stylistic register. Second, and even more important, this focus facilitates a consideration of the role played by pleasure in making natural-historical and natural-philosophical subjects intelligible. We have already seen that the comparison, understood as an epistemologically sound analogy, was an important tool in the work of producing knowledge, necessary to make invisible or otherwise imperceptible things intelligible. But the previous chapters have also shown both that those comparisons had a rhetorical dimension and that the accounts of sensation and cognition explaining how they stimulated mental images gave a large role to affect. My aim here, therefore, is to consider more fully the rhetorical, affective aspects of one of the main tools of natural-historical and natural-philosophical representation, at least as Grew, Hooke, Ray, and Willis understood it.

Willis opened the *Cerebri Anatome* with a dedication to Gilbert Sheldon (1598–1677), who was at that time the Archbishop of Canterbury. Seeking to praise his dedicatee, Willis employed a series of lofty metaphors.[30] Comparing the relatively modest room in which he had carried out his anatomies to the Temple of Solomon, he represented Sheldon as the temple's high priest. Placing himself in the scene as a supplicant, he made the human and animal corpses that he had dissected into sacrifices offered at the altar over which Sheldon presided: "Therefore after I had slain so many Victims, whole Hecatombs almost of all Animals, in the Anatomical Court, I could not have thought them rightly offered, unless they had been brought to the most holy Altar of Your Grace."[31] This comparison is appropriate to a dedication, since it amplifies its subject, inspiring ideas far greater than the things that it represents. There are further instances of such metaphors and figuration in the epistle to the reader, ending with an allusion to the story of the birth of the Roman goddess of wisdom, Minerva. She had been born, according to myth, after Jupiter sought Vulcan's help in alleviating the headache that he had been suffering since swallowing his consort, Metis, by ordering the fire god to split open his head.[32] Using this tale, Willis alluded to the need for his unsettling brain dissections: "Antiquity may not have said altogether in vain, That Minerva was born from the Brain, Vulcan with his Instruments playing the Midwife: For either by this way, viz. by Wounds and Death, by Anatomy, and a Cæsarean Birth, Truth will be brought to Light, or for ever lye hid."[33]

The rest of the work—which describes the brain and nerves and makes suggestions about how they function—is filled with comparisons written in a far lower register. Almost all those comparisons liken the forms and textures of the brain to commonplace things of a scale that is easily visible to the unassisted eye. Moreover, Willis called attention to the metaphorical status of them all by using words such as *quasi* or *velut*, almost always rendered in the English translation with the expression "as it were." The fornix of the brain, to take one example, is compared to a "string or ligament," thus: "which *Fornix* so called, or arched Vault, is as it were a string or ligament."[34] We have already encountered several comparisons like these, perhaps most strikingly in chapter 4's discussion of the images used by Grew and Ray. But it bears repeating that those two naturalists both used comparisons of the same sort when they came to describe the forms of natural bodies, also paying the same scrupulous attention to the degree of likeness between the comparison and the thing itself. Consider two examples from Grew's *Anatomy of Plants*, the first of which is his description of what happens to the layers of the skin of a bean when dried: "When 'tis dry, they cleave so closely together, that the Eye not before instructed, will judge them but one; the inner Coat [. . .] so far shrinking up, as to seem only the roughness of the outer, somewhat resembling Wafers under Maquaroons."[35] Later, in his description of the pores of the pith of plant roots, he took care to explain that their structure was only "somewhat answerable to the Cells of an Hony-Comb."[36] The style of these descriptions could be described as "low" because the objects used for comparison are commonplace, medium-scale, and easily visible.

Ray used comparisons in the same way Grew did, deploying an array of common, medium-scale objects. Take this description, from his correspondence, of a flower he called *Serpentaria virginiana*, a species in the modern genus of snakeroot (*Ageratina*) named by Linnaeus *Aristolochia serpentaria*, and its difference from another flower:

> Near the ground grow one or two hollow flowers [. . .] different in form from the *Pytolochia retica*, or any other yet known; all whose flowers [. . .] resemble a cow's horn, the top growing to the rudiment of the seed-vessel, and the open end cut slanting like a drenching-horn, whereas this of ours terminates with a heel, which supports a broad round galericulated [as if covered with a cap or hat] lip, the centre of which opens into the hollow of the flower [. . .]. The seed-vessel is hexagonal, shaped like a pear, when full-grown nearly half an inch in diameter.[37]

The comparisons in Ray's Latin works, as we have seen, have the same character—recall his vivid description of the mouth structure of the thornback ray, *Raja clavata*, with its "rude rhomboidal protuberances, resembling carpenters' files."[38] If we extract these comparisons from their context and place them alongside one another, we are confronted by a row of simple, medium-scale objects that are easy to imagine: "cow's horn," "drenching-horn," "heel," "lip," "pear," "carpenters' files." Ray did not systematically refer to any individual thing or sets of things when he made comparisons. Neither did he use things that formally resembled plants and animals in any way other than in their shapes. If there is consistency in this metaphorical tone, it is in its use of medium-scale, commonplace objects to help readers easily form images of the shapes of things.

The same quality is evident in the *Cerebri Anatome*, which deploys a wide variety of comparisons, both throughout the whole text and within representations of individual organs and systems. This can be demonstrated by exploiting Willis's consistency in indicating the metaphorical status of his comparisons. By searching Samuel Pordage's translation for all instances of the expression "as it were" (the term he almost always used in the place of Willis's *quasi* and *velut*), it is possible to catalog nearly all the comparisons. In the 66 folio pages dealing with the brain, there are 158 comparisons introduced by the phrase "as it were." They appear throughout the text, although they are most thickly deployed to describe points of difficulty or importance. This count leaves out many metaphors often to be found in anatomical nomenclature, such as "vermiform" ("wormlike").[39] Typical examples are the following:

> "a net admirably variegated or flourished"
> "Chambers or Vaults"
> "two stems"
> "two out-stretched wings"
> "a Flood-gate"
> "a bubble"
> "transverse strings or cords"
> "the serpentine chanels of an Alembick"

a "*Balneo Mariæ*" (a bain-marie), a body patterned "as it were with furrows"
"the bill of a *Pelican*"
"a Cylinder rolled about into an Orb"
"four Mole-hills"
"the Kings High-way"
a "Machine or Clock"
"distinct Store-houses"
"little Tad-stoles or Puffes"
"so many little holes in a Honey-comb"
"the Chest [. . .] of a musical Organ"[40]

There is one exception to this pattern: a comparison of the shape of the *medulla oblongata* to "the Poets *Parnassus* [. . .] like the letter Y." Even this outlier, however, finds its resolution in a straightforward reference to a letter of the alphabet.[41] In every other instance the comparison employed directs the reader to a commonplace object, of a scale that is easily visible.

This analysis reveals two things that might otherwise have remained hidden about the comparisons of the *Cerebri Anatome*. The first is that, as metaphors, they are mixed. Willis's comparisons possess neither a consistent internal scheme nor fixed relationships to the forms and textures for which they stand. Whether representing individual structures and systems or reaching for comparisons to introduce hypotheses about how they functioned, he referred to a wide range of objects. Second, despite mixing his metaphors, Willis exercised an undeniable consistency in his manner of amplifying the structures and systems of the brain. He almost exclusively used comparisons of a medium scale—generally much larger and always more easily visible than the bodies for which they stood. In fact, the only thing that his comparisons have in common is that they refer to familiar things that are easily apprehended by the senses—and recalled by the memory—because of their medium scale. Therefore, we might say that the style of the *Cerebri Anatome* is not, according to Lamy's criteria, entirely plain. It amplifies its subject, making the obscure and difficult-to-imagine structures of the brain and nerves easier to apprehend.

Ease, Pleasure, and Intelligibility

In Lamy's *Art of Speaking* there are two general discussions of the stylistic qualities that we have identified in Willis's *Cerebri Anatome*. The first comes in a section recommending different styles for different genres of writing and speech, one of which is "Natural or Moral Philosophy." There, Lamy focused largely on the ways in which those branches of philosophy differed from geometry, reminding his readers that geometry required no embellishment because of its simplicity and total lack of ambiguity. In contrast, natural and moral philosophy, requiring language for their communication, had to be far less austere. Indeed, they were capable of eliciting pleasure, or at least of not putting "the Reader into an ill humor by studying them." Therefore, Lamy argued, a small amount of rhetorical embellishment was appropriate.[42] This recommendation is worth pausing over because of its ambiguous treatment of the pleasures and pains of philosophical discourse. There seems to be another reminder here that Ray was right to worry that his readers might not enjoy the study of nature. Regardless, the *Art of Speaking* endorses Willis's exact stylistic approach. At several points in the text, Lamy encouraged would-be orators to use metaphors referring to things that their readers would encounter in day-to-day life, explaining in one place that "the best way will be to take our Metaphors from sensible things, and such as are frequently represented to our Eyes, whose Images are easily apprehended without scrutiny or trouble."[43]

Perhaps even more tellingly, Willis's stylistic register is also closely reflected in the suggestion for a reformed rhetoric laid out—as we saw in chapter 4—by Thomas Sprat in his *History of the Royal Society of London for the Improving of Natural Knowledge*. Unlike the *Art of Speaking*, Sprat's text contains explicit recognition that many contemporary readers were not likely to find such simple metaphors very pleasurable. The *History* attempts to persuade readers that comparisons to simple, medium-scale bodies could form the foundation of a reformed rhetorical taste combining the pleasures of rhetoric with an adherence to edifying truths. Sprat urged his readers to employ a new kind of rhetoric that would furnish a "vast treasure of admirable Imaginations," while doing nothing to distort the truth or nourish false pleasures. Those were comparisons to medium-scale things, easily apprehended by the senses—the products of the manual and natural arts:

> The Comparisons which these [the manual arts] may afford will be intelligible to all, becaus they proceed from things that enter into all mens Senses. These will make the most vigorous impressions on mens Fancies, becaus they do even touch their Eyes, and are neerest to their Nature. Of these the variety will be infinit; for the particulars are so, from whence they may be deduc'd:[44]

In a passage thick with the language of *enargeia*, Sprat here hinted at the possible power of a reformed, virtuous rhetoric that sought out comparisons in the world of things that may easily affect the senses. By his own confession, he was trying to get people to enjoy a rhetorical register that was not customarily associated with much pleasure. He therefore tried to persuade readers that such comparisons could lead to gratification, and even prove more useful for the purposes of oratory than those that were more florid, because of their address to the senses.

But to work out what sorts of effects Willis wanted to provoke with his comparisons, we need to give some more thought to how his style related to the demands of the subject and audience. In this regard, the second of the stylistic discussions in Lamy's *Art of Speaking* relevant to interpreting the style of the *Cerebri Anatome* proves useful. It comes in a section discussing metaphors that either were mixed or otherwise had disproportionate relationship to their subjects. In some cases, Lamy explained, the use of such metaphors was a breach of decorum. In other cases, however, he pointed out that those metaphors were permissible. Lamy resolved this apparent contradiction by invoking the discourse of accommodation, arguing that what counted as a stylistic failure in one constellation of subject, speaker, and audience might very well count as a great success in one that was differently configured. In general, he argued, mixed metaphors with little formal resemblance to their subjects were to be avoided, especially if their purpose was solely ornamental.[45] He made an exception, however, in cases where the subject at hand was particularly difficult for readers to imagine—for example, in the description of the passions displayed by the participants in heroic combats and other things lying outside daily experience. In cases such as these, the needs of intelligibility and pleasure were better served by metaphors that would usually be accounted disproportionate. Turning to an example drawn from Virgil's *Aeneid* to make his point, Lamy argued that its apparent breaches of decorum were in fact justified by the need to accommodate certain actions to the capacities of the audience. Where intelligibility was at stake, it was permissible to mix metaphors.

The example comes from the episode in which Camilla, leader of the Volsci, slew a young Ligurian in battle. Virgil compared the action to that of a hawk swooping on a dove and scoring an easy meal, adding at the end for emphasis, "The feathers, foul with blood, come tumbling to the ground."[46] There were two possible objections to the comparison. First, it contains elements that do not cohere with the action for which they stand: there are no falling objects to which the bloody feathers might be compared. Second, the comparison is in a rather low register, ill fitting for so elevated an action because it refers to a common bird. Lamy, however, made the point that a combat such as Virgil described lay outside the bounds of normal experience, meaning that few people possessed the requisite store of mental images to adequately conceive it. This circumstance made a virtue of the flawed metaphor. The inclusion of the falling feathers, for example, was to be applauded because it helped render "a more sensible description of a Pigeon torn in pieces by a

Hawk" than could have been supplied by the metaphor alone. But the lowness of the metaphor was its most important aspect—the aspect most likely to aid intelligibility and even to induce pleasure. Lamy suggested that it might have been hard mental work to imagine the succession of fantastical scenes described in the *Aeneid*. The familiarity of the image of the hawk and the pigeon, along with its appeal to the senses, helped mitigate the bombardment of so much lofty imagery: "He does it to ease and relax the Mind of the Reader, whom the Grandeur and Dignity of his Matter had held in too strong an intention."[47]

Lamy's defense of Virgil's mixed metaphors perhaps gives us a clearer sense of the kind of pleasure to be had from the comparisons in the *Cerebri Anatome*, along with their epistemological utility. By continually comparing the anatomy of the brain and nerves to an array of commonplace, medium-scale things, Willis sought to mitigate the difficulty of imagining the hidden recesses of the brain and nerves, which under normal circumstances were visible only to skilled anatomists. This difficulty was the basic condition of natural history. Willis alluded to it throughout the epistle to the reader, explaining that he had rejected the "easie" path of dreaming up a "Poetical Philosophy and Physick neatly wrought with Novity and Conjectures" in favor of the hard work of anatomizing and describing the brain and nerves in the hope of working out their functions.[48] And yet Willis represented the results of his painstaking investigations using a style designed to enable readers to make easy mental associations between the bodies being described and the memories of day-to-day objects. Willis's approach to representation was consistent with what we learned in chapter 4 from the letters sent between Ray and his correspondents. Those letters testify at one and the same time to the hard work that went into investigating and representing the designs of natural things, and to the fact that Ray and his friends wanted their representations to work with as much ease and pleasure as possible.

This division of labor also recalls once again Locke's distinction between the mental work of judgment and the easy, pleasurable operations of wit. Investigating and representing the designs of natural things was difficult work. This work, with its emphasis on making minute distinctions between similar things, closely resembled Locke's description of the difficult, unpleasant mental operations of judgment. Yet the representations resulting from this exercise were intended to work in the opposite fashion, provoking the mind into reaching for easy associations between different images. According to Locke, this was the mental process that made "wit" pleasurable. Indeed, he asserted that poetry and rhetoric owed their capacity to inspire pleasure to the fact that the mind experienced pleasure when it easily identified similarities between different things. I do not want to suggest that every contemporary reader would have felt pleasure when reading the *Cerebri Anatome*. Indeed, Steven Shapin's essay on attitudes toward natural philosophy in fashionable circles of the seventeenth and eighteenth centuries shows that Ray's encouragements to enjoy the study of nature often fell on deaf ears.[49] It is important, however, to recognize that Ray and others saw their work as an exercise in the cultivation of taste—an effort to impart the capacity to enjoy natural histories as much as to understand them. Willis's aim was to overcome the difficulty of imagining the intricate forms of the brain and nerves by means of the easy and pleasurable mental picturing of familiar objects.

The Bodily Pleasures of the Imagination

Taking Willis's account of sensation and imagination into account, we can see that the pleasures flowing from his descriptive style were—at least in part—pleasures of the body. Since we dealt with Willis's hypothesis in chapter 2, let us here recapitulate just the most crucial aspects of his account of ideation. The faculties of the memory and imagination are here particularly important because the comparisons of the *Cerebri Anatome* depended on the reader being able to recall images that had earlier been impressed into the memory resulting from sensation and to recombine them using the faculty of the imagination. According to Willis's hypothesis, mental images arose when the impressions left by external things on the senses were transmitted through the brain's spirits to the cerebrum, the seat of the imagination, and the cortex, where memories were deposited. Crucially, as we saw, the likelihood that an impression would become an idea or a memory depended on the force with which that impression was transmitted through the brain's spirits and fibers.[50] While Grew, Hooke, and Ray disagreed with Willis about the extent to which the operations of the "sensitive soul" could be accounted for by these mechanical operations, they nevertheless agreed with his fundamental claim that all or most ideas came from impressions made on the brain and senses, and that it was not in the imagination's capacity to form absolutely new ideas, but only to compare and reorder images laid up in the memory.[51]

In the companion text to his anatomy, *De Anima Brutorum*, Willis hypothesized that particularly long-lasting and vivid memories were caused by the conveyance of forceful and repeated impressions to the *cortex*.[52] While Robert Hooke, in the seventh of his "Lectures of Light," proposed entirely different (and much more speculative) mechanisms for the formation of ideas and memories, he nevertheless agreed that forceful or frequent impressions on the fabric of the brain led to vivid memories:

> The Memory of things long since done is for the most part very faint, unless in some cases, where the Impressions made upon these Ideas were at first very powerful, or often recalled, which may be said to be a new forming of them.[53]

We can see, then, that the vivid, pleasurable quality of the comparisons employed by Grew, Ray, and Willis had a physiological explanation. Comparisons to "drenching horns," "carpenters' files," and the "little holes in a Honeycomb" were effective because the medium-scale objects that they referred to were of a scale proportionate to the organs of sensation and cognition. Consequently, they had been inscribed easily, forcefully, and often into the memory, making them easy to recall and recombine to form images corresponding to the natural body being described.

Willis's *Cerebri Anatome* therefore furnished powerful arguments for the style in which it had been written. Comparisons to familiar things were likely to stir the memory to easily and pleasantly recall images that had been forcefully and repeatedly impressed into its fabric. In his own attempt to hypothesize mechanisms for the mechanics of cognition, Hooke made this point explicitly. At an early point in the lecture he explained that, to make his hypothetical cognitive mechanisms easier to understand, he would employ comparisons to things that may easily be sensed:

> Now because nothing is so well understood or apprehended, as when it is represented under some sensible Form, I would, to make my Notion the more conceivable, make a mechanical and sensible Figure and Picture thereof, and from that shew how I conceive all the Actions and Operations of the Soul as Apprehending, Remembring and Reasoning are performed.[54]

Hooke likened the cumulative effect of his rhetoric to a form of picturing, suggesting that his words would lead the mind to form a "sensible Figure and Picture" of the things that he was seeking to represent. Like Willis, Hooke employed comparisons to provoke the memory of things that had left—or had the capacity to leave—deep impressions in the fabric of the brain, stirring up vivid and potentially pleasurable mental images.

The Cultivation of the Brain, Nerves, and Senses

The physiological basis for the pleasures of rhetoric laid out in the *Cerebri Anatome* has important implications for our understanding of how natural history and anatomy may have contributed to the project to enable readers to enjoy the study of nature. Willis attempted to show in both the *Cerebri Anatome* and *De Anima Brutorum* that the generation of mental images and memories depended on the spirits and fabrics of the brain being healthy. If the brain was somehow unfit to receive impressions and combine images, the mind would be capable neither of rightly apprehending the objects of sensation nor of producing new combinations of ideas. At the same time, as is clear from the use of the word *impression*, Willis suggested that the brain was materially changed by its encounters with powerful sensations. It was therefore possible, in other words, for powerful rhetoric to alter the fabric of the brain and nerves. In this literal, bodily sense, the rhetoric of natural history and anatomy could be an agent in the cultivation of the bodily parts of the mind. Lamy made this neurological inflection of rhetoric explicit in a section of the *Art of Speaking* with the title, "The Qualities of the Substance of the Brain, and the Animal Spirits, are necessary to make a good Imagination." He wanted to show that the successful impression of ideas in the brain depended on the consistency of its fibers:

> Figures drawn upon the Surface of the Water leave no prints, because they are immediatly filled up. Figures ingrav'd upon Marble are seldom perfect, because the hardness of the Matter gives too much resistance to the Chissel. This gives us to understand that the substance of the Brain ought to have certain Qualities, without which it cannot receive exactly the Images of such things as the Soul imagines. If the Brain be too moist, and the little Threads and Fibers which compose it too feeble and lax, they cannot retain the Foldings and Impressions given them by the Animal Spirits, and by consequence the things drawn there are confused, and like those we endeavour to draw upon Mud: If the Brain be too dry, and the Fibers too hard, 'tis impossible all the strokes of the Objects should leave their Impressions, which makes every thing seem dry and meagre to men of that Temper.[55]

With this description (and take note of the comparisons to water, marble, and mud), Lamy accounted in bodily terms for the fact that some people responded to the impressions of rhetoric in different ways. His use of "Temper" here is telling. Lamy was thinking about the matter in medical terms, linking the humoral temperaments of medical theory (bilious, melancholic, phlegmatic, and sanguine) to the makeup of individual brains.

It followed that sufficiently powerful rhetoric was understood to work its effects by altering the fabric of the brain and sensory organs, changing the way they perceived, stored, and ranged ideas. Lamy made this point, while outlining its consequences for speakers and writers, in a section entitled "The Advantage of a good Memory." The brain had a tendency, as he saw it, to lose the capacity to register and range ideas unless it was regularly exercised. The problem was that the "Fibres of the Brain do harden and grow stiff, if that stiffness be not prevented by frequent folding them." The solution to this problem, therefore, was to continually expose and reexpose the brain to appropriate sensory stimuli. Following such a regime, he suggested, would maintain the flexibility of the fibers. But it also raised another difficulty. If the brain were exposed to texts written in a poor style—especially a style that corrupted the relationship between words and the things they stood for—the fibers could end up being shaped in such a way as to damage the brain's capacity to apprehend the truth or even to produce truthful notions. On this score, Lamy made his case by comparing the actions of stimuli on the senses, and their remaining impressions in the memory, to those of mechanical type on paper. Even if the subject of the discourse were itself wholesome, it would exercise a damaging effect on the mind if it were clothed with badly chosen words:

> An excellent Person has resembled the Memory to a Printing-Press; a Printer who has none but Gothick Characters, prints nothing but in Gothick Characters, let the Treatise be never so good. The same may be said of those whose Memories are ful of nothing but improper words; having nothing in their minds but Gothick Molds, and their thoughts clothing themselves with Expressions from thence, no wonder if they always assume a Gothick aire and fashion.[56]

Failing to dress a discourse in an appropriate style, a writer could end up producing effects other than those intended. No matter how good the subject at hand was, a poor style would render it less forceful, and perhaps even confusing. Even worse, that style might muddle the very fabric of the brains of those people it was intended to help. To habituate oneself to any powerful sensory stimuli was—this cannot be emphasized enough—to alter the contexture of the brain and nerves.

When Lamy implied in this passage that the opposite effects were also possible, he meant it literally. The health of the brain could be improved by exposing it to good texts written in good styles. The rhetoric clothing virtuous texts could, if appropriately executed, improve the brain's capacity to easily register and range ideas. However, the skill of the author was not the only factor at play. Readers had to be willing to help themselves by habituating themselves to those texts. Indeed, they could only cure the material parts of the mind by repeatedly attending to them. Those who did so had the chance to improve their capacity to apprehend ideas about the world around them, to distinguish truth from falsehood, and to make good judgments about a range of things. The model of mental cultivation that Lamy proposed was therefore one that saw inherent connections between the production of subjective states of affect and the successful acquisition of knowledge. Such a model can help us explain how it was that Ray could claim that readers needed to possess highly particular subjectivities if they were to acquire knowledge of nature. At the same time, I would suggest, it helps us see how their representations of design in nature could meaningfully serve as instruments in the prescription of those subjectivities, and their transmission from naturalists to more general readers. The style of the *Cerebri Anatome* was such an instrument—a means both to transmit Willis's judgments about the forms and purposes of the brain and nerves and to engender pleasant clarity in the minds of readers.

Many of Willis's contemporaries bought into this embodied model of habituation. Adrian Johns has shown, for instance, that medics and natural philosophers of the seventeenth century took a close interest in the "physiology of reading"—the notion that acts of reading could either cure or distemper the parts of the mind dependent on the body. Indeed, such ideas often featured in treatises on nervous medicine and pedagogy, including the pathological part of Willis's *De Anima Brutorum*, where Willis went so far as to warn that parents who read excessively could so weaken the animal spirits that they might pass "stupidity" on to their children.[57] As Johns and others have also explained, Boyle himself experienced the unhealthy effects of misapplied reading in his youth, when physicians prescribed a regimen of reading—the "stale Adventures [of] Amadis de Gaule; & other ~~Raving Bookes~~ Fabulous & wandring Stories"—to cure him of melancholy. This course of romantic fiction only caused Boyle more harm, however, weakening his imagination. While Boyle subsequently calmed his imagination by pursuing a regimen of mathematical exercises, he never believed himself to have been entirely cured, finding himself still subject to the "Habitude of Raving" that he picked up in his youth.[58] Boyle's preoccupation with the physiology of reading should prompt us to take seriously the gustatory and medical metaphors he used in the *Style of the Scriptures* to represent the process through which a reader could learn to enjoy the style of the Bible. The "distemper'd Palate," he wrote, could be reformed by the consumption of "wholesome Aliments." Although Boyle did not make the point explicitly in that text, both he and his contemporaries might well have understood those metaphors in literal terms. To be sure, the act of reading required the guidance of the rational, immaterial soul. At the same time, however, it could only be successfully undertaken by learning to manage the affective, embodied processes on which that soul depended for almost all its ideas. For Boyle, the management of somatic pleasures and pains was central to the acquisition of moral and intellectual benefits from books and other objects of sensory experience.[59]

In addition, there is a growing body of evidence indicating that many thinkers of the sixteenth and seventeenth centuries regarded the practice of philosophy, including the study of nature, as a form of medicine for the mind—that is, a form of habituation that would not only make individuals better able to discover the truth but also give them a greater capacity for virtue. Sorana Corneanu has shown, for instance, that Locke regarded the act of paying attention to a given object as a passionate act—an act that sprang as much from the effects of pleasure and pain on the will as it did on the dictates of reason. As Locke explained in his essay *Of the Conduct of the Understanding* (1706), the successful exercise of reason depended in large measure on habituating the passions to enjoy the objects that would lead the mind closer to the discovery of the truth. For Locke, the aims of philosophy could only be accomplished through careful management of the body's capacity for experiencing pleasure and pain in its encounters with external objects.[60] Several other scholars, including not only Corneanu herself but also Jan Golinski, Lisa Jardine, Steven Shapin, and Charles T. Wolfe, have demonstrated just how much the sort of habituation described by Locke depended on embodied practices drawn from dietetics and medicine. Golinski has established, for instance, that philosophers including Boyle, Locke, and Newton all pursued therapeutic regimes that would make their bodies more effective vehicles for the pursuit of knowledge. Meanwhile, Jardine has used Hooke's diary to reveal that Hooke attempted to enhance his cognitive abilities through a course of self-experimentation with drugs, including powerful purgatives. Ironically, Hooke eventually habituated his body to those harmful substances, requiring bigger and bigger doses to obtain the state of mental clarity he desired.[61]

I would therefore suggest that the process of habituation described by Ray in *The Wisdom of God* therefore had a bodily component. I would add, moreover, that the descriptive style he and his contemporaries used was intended to be an agent in that process. In fact, Ray brought together the same combination of stylistic and neuropathological concerns in his own discussions of the effects of reading inappropriate books. Embellishment included for its own sake, Ray wrote, was "apt to *infect* a man with such *odd Humors* of Pride, and Affectation, and Curiosity, as will render him unfit for any great Employment."[62] Like Boyle, he worried that gratuitously pleasing rhetoric could harm the constitution of the mind, provoking a disposition to affective states that would in turn prejudice the pursuit of both knowledge and virtue. It follows, likewise, that Ray thought it possible to heal the cognitive and affective faculties of the brain by exposing it, step-by-step, to virtuous and truthful things. This is the sense that we get, at least, by comparing Ray's project for the cultivation of the mind through the study of nature with Boyle's cure based on regular Bible reading. He and many of his contemporaries thought that the fabric of the brain could be altered by reading and other sensory acts. They also agreed that the material parts of the mind could be cured by habituating them to true and virtuous forms of rhetoric that, by their address to the senses, could work powerful effects on the fabric of the brain. Those effects included an enhancement of the cognitive faculties themselves and an alteration of the affective disposition. The impressions left on the brain by the style of natural philosophy made the acquisition of true judgments about the world easier and more pleasurable.

Aesthetic Science

This book opened by posing questions about how the natural philosophers of the early Royal Society responded to the dilemmas thrown up by the desire to use the senses as instruments in the production of putatively objective knowledge. There, I remarked that scholars have for the most part emphasized how natural philosophers sought to guarantee objectivity by somehow disciplining their own and other people's affective dispositions.[63] The account of representation offered here confirms this view—at least in its broad outlines—in two respects. Undoubtedly Ray, Boyle, Grew, Hooke, and Willis did in fact aim to discipline the experiences that their readers would have of nature. And they did so, moreover, to guarantee that those readers would acquire ideas corresponding to the ones that had been intended. In this chapter and chapter 4, however, I have shown that some of the Royal Society's leading members thought about the mechanisms by which the transmission of knowledge was to be guaranteed in an entirely different manner than has been suggested. When they represented plant and animal bodies in printed books, they did not seek to make the body's responses to sensory stimuli dispassionate. Instead, they saw their representations as tools in the transmission of appropriate passionate responses—those accompanying the easy and pleasurable acquisition of knowledge through the organs of sensation and cognition.

As we saw in the previous section, Sorana Corneanu and others have weighed in against the argument that the aim of late seventeenth-century empiricism was to efface individual subjectivities through the rigors of philosophical method, new forms of representation, and strategies for collective witnessing. They have shown, by contrast, that philosophers such as Francis Bacon, Boyle, and Locke saw their epistemological programs as disciplines for habituating individuals' minds and bodies for the exercise of good judgment. The key difficulty with this model of habituation, however, is that it does not explain how natural philosophers sought to communicate knowledge. It does not explain how they sought to transmit observations from one member of the philosophical community to another. In this chapter, I have sought out the beginnings of an answer to this crucial question by focusing on the strategies natural philosophers used to guarantee that their readers acquired the same habits and affective responses they sought from themselves. The discourse of taste provides us, I would suggest, with a powerful conceptual vocabulary for describing the strategies used by naturalists as they sought to generate consensus about the meanings of experience. It is worth pointing out, moreover, that they also mobilized this discourse themselves, explicitly framing natural philosophy as an exercise in the cultivation of the capacity to take pleasure in something that might not seem enjoyable at first glance. Although Ray and his contemporaries sought something like objectivity, in the sense that they wanted to produce facts that would command universal assent, they did not align that objectivity with a disembodied or dispassionate model of experience. They were too keenly aware of the potential effects of the body's habits on its perceptions to think like that. Instead they posited, to use modern-day language, a bodily kind of intersubjectivity, or—to speak using the language of Kant's *Critique*—a subjective universal. The communicability of knowledge was to be guaranteed not only by the vividness of its representation but also by the effects of that representation on the bodies and thence the affective dispositions of readers.

Finally, this chapter's focus on the affective work of verbal representations has made it possible to complete our reconsideration of the relationship between physico-theology and the main branches of natural philosophy with which it was mostly concerned—anatomy and natural history. The moral and theological betterment promised by Ray in *The Wisdom of God* was built into the knowledge-making fabric of natural philosophy. If attended to, its representations could effect the bodily and mental transformations that would permit readers to successfully enjoy the evidence of God's activity in the world. In chapter 3, I argued that the work of representing natural bodies was as much concerned with the transmission of judgments about their design as it was with the provision of bare knowledge of their forms. In this chapter, however, I have gone a step further by showing that the representations of the *Cerebri Anatome* and other texts were also instruments in the transmission of ideal affective states, understood literally as the disposition of the corporeal parts of the mind to receive certain kinds of impressions and to take pleasure in them. The striking thing about the affective register of the *Cerebri Anatome* is that it closely resembles the one described and encouraged by Ray in *The Wisdom of God*. Indeed, when Ray tried to persuade his readers that the study of nature could become "easie and familiar" if they only paid it enough attention, he might as well have been describing the effects on the brain envisaged by Willis when he compared the complex structures of the brain and nerves to wings, bubbles, and honeycombs. More generally, the temperate delights described in physico-theology are reflected in the air of pleasant instruction identifiable in the style of the *Cerebri Anatome* and other natural-historical and anatomical representations.

It is therefore difficult to sustain the claim that—because of its use of affective rhetoric—physico-theology belonged to a realm of persuasion that was entirely distinct from the sober, rational work of natural history. Two things have instead become clear, both suggesting that the affective, epistemological, moral, and theological claims made by the physico-theologians were closely intertwined. First, the provocation of affective states was important to the knowledge-making work of natural history and anatomy, serving as a guarantee that ideas corresponding to those observed and imagined by the naturalist would be reproduced in the minds of individual readers. The capacity to make good judgments about nature could in some cases depend on the cultivation of a certain kind of subjectivity—what we might call the cultivation of a taste for the study of nature. Second, it is likely that Grew, Hooke, Ray, and Willis saw the acquisition of knowledge through their natural histories as a form of therapy for the bodily parts of the mind. The reward for paying attention to representations of nature was participation in an active process of regeneration, one that would result in an improvement in the mind's capacity to judge the world and its capacity to enjoy the act of attention on which this was all premised. The program for moral regeneration promised in physico-theology therefore took substantial, corporeal form in the mental transformations accompanying the attentive enjoyment of descriptions and drawings of plants and animals.

Conclusion

Embodied Aesthetics

Taste has an ambiguous place in Western thought. The philosopher Giorgio Agamben remarks that the term is frequently used to denote the lowest of the human senses, denoting a gustatory faculty that reminds us of our close connection to animals. At the same time, however, it also refers to one of the highest of human accomplishments—what Agamben calls the "special form of knowledge that enjoys the beautiful object and the special form of pleasure that judges beauty."[1] Scholars have so far paid little attention to the links between the empirical sciences and the development of practices for forming judgments of taste, whether understood in either of these two senses. Understood as a form of judgment somehow exercised by the tongue, taste is difficult to analyze. As Steven Shapin has pointed out, such judgments seem to emerge from processes of embodied intuition that resist the analytical methods used by those seeking to trace the emergence of scientific objectivity.[2] A similar point, meanwhile, can be made about taste when it is used metaphorically to describe judgments about the beauty of art objects. As we saw in the introduction, thinkers such as Immanuel Kant argued that such aesthetic judgments emerge from cognitive operations distinct from those at stake in the discovery of moral or scientific truths. As a result, there has been a tendency to overlook the role of taste in the history of science. Although considerable attention has been paid to the links between the arts and sciences, very little of that scholarship calls into question the Kantian distinction between the production of intersubjective agreement about matters of taste and the production of what passes for objectivity.[3]

In the Royal Society of the seventeenth century, however, this distinction—framed by Kant in explicit opposition to the aesthetic and epistemological claims of physico-theology—was far less meaningful. Philosophers such as

Robert Boyle, Nehemiah Grew, Robert Hooke, John Locke, John Ray, and Thomas Willis gave pleasure an ambiguous status in their philosophical and theological projects. They all worried that feelings of pleasure could sometimes prejudice the mind against the truth. Moreover, they all recognized one way or another that pleasure did not necessarily arise from the part of the mind that they held responsible for the exercise of reason. Ray and his contemporaries did not, however, deal with the affective dimensions of experience by seeking to exclude them from the discovery of purposiveness in nature. On the contrary, they identified close links between the capacity to experience specific kinds of pleasure and the ability to discover the truth about God's involvement with the physical world. The belief that the world, including human responses to sensory experience, was the product of divine design gave this alignment between pleasure and the production of knowledge an ontological foundation. Recognizing that humans did not all experience the world in the same way, Ray and his contemporaries interpreted the inability to take pleasure in divine design as a product of human corruption. As a result, they aimed not only to provoke the pleasures arising from the encounter with specimens of divine design but also to give others the capacity to experience those affective states for themselves. Experienced as a form of bodily and mental cultivation, this aesthetic science would eventually give its practitioners the capacity to take pleasure in things that would do them good, yielding cognitive and moral benefits. Ray and his contemporaries did not, moreover, draw sharp distinctions between the pleasures of the mind and those of the body. Recognizing the mind's dependence on bodily organs for its perceptions of the external world, they sought to mobilize the desires and pleasures of the body in the service of intertwined moral and epistemological objectives.

To get a sense of how this insight might change how we view the empirical projects of the seventeenth and eighteenth centuries, let us turn one last time to the single most influential statement of the idea that human knowledge comes from experience—Locke's *Essay Concerning Human Understanding*. Locke devoted only a small portion of the *Essay* to free will and ethical questions. Yet his brief attempt to outline a system of ethics coherent with the idea that humans get most of their ideas from sensory experience led him into difficulties. In the chapter dealing with free will, Locke argued that human behavior was motivated by pleasure and pain, or at least by the expectation that those experiences would arise from a given action. Defining human action this way, he made good and evil into relative terms, characterizing them not as moral absolutes but rather as subjective, individual perceptions. Evil was nothing more than a person's sense of what would bring pain, while good was whatever object or act seemed likely to cause pleasure. Rejecting the notion that humans were born with innate ideas, Locke therefore made morality into a matter of convention.[4] At the same time, however, he wanted to show that the world should be governed by binding moral laws flowing from nature—precepts that any reasonable person could be expected to figure out and follow. Locke therefore suggested that individuals could work out the best course of action by making an accurate assessment of what was most likely to bring them lasting happiness. There was no danger, he suggested, that anybody capable of accurately weighing up the infinite pleasures of heaven against the transient pleasures available to mortals would choose a life of vice over one of virtue. Many interpretations of Locke's ethics therefore identify it as a hollow appeal to rational self-interest. It is certainly true that he ran the risk of reducing the good life to a rational calculus that placed the pain resulting from bad behavior in the balance against the pleasure to be gained by behaving well.[5]

Locke was by no means confident, however, that reason alone could make people endure present pains to secure future pleasures. As we have already seen, Locke took a close interest in the possibility that the rapid and pleasurable formation of vivid mental impressions could overpower the labored discriminations of rational judgment. It is thus a bit surprising that little attention has so far been paid to Locke's frank admission that the experience of pleasure would play a crucial role in giving people appropriate moral dispositions. Indeed, he dwelled at length on the subjectivity of pleasure and pain, explaining how individuals appeared to experience them in dramatically different—and mistaken—ways.[6] Rather than seeking to subject those fickle feelings directly to reason, however, he attempted to show how they could be made to serve as the foundation for a stable system of morality. His solution was to make the pursuit of the good life into an exercise in cultivating a capacity for the right kinds of pleasure:

> The last enquiry therefore concerning this matter is, Whether it be in a Man's power to change the pleasantness, and unpleasantness, that accompanies any sort of action? and to that, it is plain in many cases he can. Men may and should correct their palates, and give a relish to what either has, or they suppose has none. [. . .] 'tis a mistake to think, that Men cannot change the displeasingness, or indifferency, that is in actions, into pleasure and desire, if they will do but what is in their power.[7]

In the *Essay* itself, Locke had little to say about the exact steps individuals might take to rectify their perceptions. He made it clear, however, that he envisaged a process of habituation, with individuals gradually bringing themselves to enjoy things they had once found unpleasant. He thus wrote that "Habits have powerful charms, and put [. . .] strong attractions of easiness and pleasure into what we accustom our selves to." The cultivation of taste and the cure of the irrational affects, it turns out, was Locke's solution to the morally and epistemologically significant problem of getting people to agree about their perceptions of pleasure and pain. Like the physico-theologians, he used the language of habit and regimen: the gradual rectification of the palate ("give a relish") enabled him to describe how to forge intersubjective agreement from the subjectivities of individual experience.[8]

We should pay more attention to the bodily and medical terms in which both Locke and the physico-theologians discussed the production of consensus about the affective dimensions of experience. Far from being mere metaphors, their discussions of habituation and regimen hold important clues about what it meant to premise both knowledge and ethics on sensory experience. As we saw in chapter 5, it is increasingly clear that thinkers of the seventeenth century integrated into the study of nature therapeutic regimes intended to make their bodies fitter instruments for the discovery of the truth. We have seen, moreover, that such regimes, intertwining rhetoric, poetics, and the neurosciences, defined the forms of embodiment and consequently the affective states appropriate to the collective production of knowledge. To conclude, I would like to suggest that these insights may enable us to find new ways of linking the history of the empirical sciences to the history of the other disciplines concerned with the affective dimensions of experience. We are familiar today with the argument, made by historians and by sociologists of science, that the history of objectivity should be understood as the story of the changing ways in which communities of scientific practitioners established procedures for reaching intersubjective agreement. Quite rightly, scholars have for the most part framed those stories in political terms, revealing that the production of knowledge was bound up with the structures of power and the conventions that articulated them. In the eighteenth century, however, I suggest that practitioners of both the arts and the sciences wanted to form communities bound by an intersubjectivity of a far more literal kind. For aesthetic theorists, art critics, natural philosophers, and medics alike, medical practices for modifying the desires and impulses of the body provided a crucial resource for obtaining consensus about the pleasures and pains of sensory experience. To understand how this could be the case, we need to link the history of the empirical sciences and medicine to the history of the other disciplines concerned above all else with the affective dimensions of experience—aesthetics and art criticism. This is the intellectual agenda I would like to sketch in the concluding pages of this book.

Disembodiment and Aesthetic Experience

The history of aesthetics, as Simon Grote has recently noted, has long been dominated by philosophers interested above all in piecing together the conceptual origins of Kant's *Critique of the Power of Judgment.*[9] Consequently, they have frequently mobilized Kantian categories to analyze earlier works of aesthetic theory, focusing on ideas that seem to anticipate Kant's definition of aesthetic experience as a form of pleasure arising from the free play of the cognitive faculties, constrained neither by the impulses of bodily desire nor by the rational determination of concepts. In the 1960s, for instance, Jerome Stolnitz argued that the modern account of aesthetic autonomy had its origins in a concept that he called "aesthetic disinterestedness"—the notion that the experience of beauty should in no way be linked to the enjoyment or anticipation of either bodily or material advantage. In turn, he suggested that Anthony Ashley Cooper, Third Earl of Shaftesbury (1671–1713) and Joseph Addison, closely followed by Francis Hutcheson, were the first thinkers to see disinterestedness as a prerequisite for the experience of beauty. For Stolnitz, the British and Irish aesthetic theorists of the early eighteenth century therefore foreshadowed Kant's insistence that the cognitive faculties needed to be free from the determination of any kind of purposiveness.[10] Although Stolnitz's argument has come in for a good deal of criticism, philosophers still tend to interpret Shaftesbury, Addison, and Hutcheson in the same broadly Kantian terms. Albeit employing different methods and exploring different ideas, scholars including Dabney Townsend, M. H. Abrams, and Paul Guyer have all interpreted the works of this generation of British and Irish aesthetic theorists through concepts and questions that would find their clearest expression in Kant's *Critique of the Power of Judgment.*[11]

In some respects, the Kantian analysis holds true. It is certainly the case that Shaftesbury, Hutcheson, and Addison made steps toward severing the experience of beauty from its sources in material things and embodied processes. All three recognized that aesthetic experience had an affective component, its pleasures arising—much like the pleasures and pains of the body—without the apparent intervention of the intellect. At the same time, their desire to align the pleasures of the sublime and the beautiful with the pursuit of virtue made it necessary to show that those pleasures could be distinguished from the forms of sexual and gustatory gratification they so closely resembled. As is well known, Shaftesbury and Hutcheson responded to this challenge by claiming that the mind possessed internal faculties analogous to, but ontologically distinct from, the bodily organs of external sensory perception. In his *Inquiry into the Original of Our Ideas of Beauty and Virtue* (1725),

Hutcheson thus equipped humans with a "sense of beauty"—a faculty of the mind that experienced pleasure upon encountering beautiful things.[12] Addison had made a similar suggestion in his famous essays of 1712 on the "Pleasures of the Imagination." There, he proposed that the arts of painting, poetry, and sculpture led neither to intellectual nor bodily gratification but instead to the pleasures of the imagination. To make his case, Addison drew on Locke's distinction, discussed in chapter 4, between the rapid and pleasurable operations of wit and the painful exercise of judgment. Depicting the imagination as an affective faculty, he contrasted its capacity for easy gratification with the hard work of the intellect. At the same time, he maintained that the pleasures of the imagination were more refined—and therefore more innocent—than the bodily pleasures to which they bore such obvious comparison.[13] Aiming to show that the enjoyment of beauty was compatible with a life of virtue, Shaftesbury, Hutcheson, and Addison all found ways to put ontological distance between aesthetic experience and the coarser feelings of the body.

The problem with focusing on this point, however, is that it obscures continuities between these putatively disembodied models of experience and the more obviously embodied ones that came both before and after, not only in medicine and natural philosophy but also in art criticism and aesthetics. From the 1740s onward, to take just one example, philosophers such as Edmund Burke (1729/30–1797) and Denis Diderot (1713–1784), along with the painter William Hogarth (1697–1764), promoted materialist aesthetic theories, arguing that the beautiful and the sublime were simply modalities of bodily pleasure and pain. For the most part, however, scholars argue that these theories marked a sharp break with the past. Aris Sarafianos, for instance, has shown that Burke drew on new vitalist theories about the role of pain in the animal economy—chiefly the experimental attempts of Richard Brocklesby (1722–1797) to validate the distinction made by Albrecht von Haller (1708–1777) between sensibility and irritability—to give a decidedly embodied account of the sublime. Rejecting Addison's suggestion that the sublime was a pleasant feeling resulting from the encounter with things too large for the imagination to grasp in one go, Burke recast it as a beneficial species of bodily pain, arising from the physical strain placed by vast objects on the organs of sensation.[14] Nobody would deny that there were important differences between the gentle and somewhat disembodied model of experience put forward by thinkers such as Addison and the thoroughgoing materialism of Burke and Hogarth. It would be mistaken, however, to insist too much on this difference. Focusing too much on how Addison, Hutcheson, and Shaftesbury sought to disembody the experience of beauty, we run the risk of overlooking their engagements with philosophical and medical discourses about the pleasures and pains of the body. Rather than representing a new and radical departure, as Sarafianos suggests, Burke's interest in the physiology of pleasure and pain might instead reflect continuity with the models of aesthetic experience developed earlier in the century.

It is worth pointing out, moreover, that twenty-first-century historians have challenged the old consensus that the aesthetic theorists of the early eighteenth century largely rejected natural philosophy as a form of dry pedantry. Brian Cowan has shown that Addison was far more sympathetic to the virtuoso culture associated with the early Royal Society than has so far been recognized. Craig Ashley Hanson, meanwhile, has pointed out that medics and natural philosophers, such as the wealthy physician Richard Mead (1673–1754), were crucial to the development of the London art market, not only as collectors but also as critics and patrons.[15] Cowan and Hanson have thus revealed that medics and natural philosophers exercised considerable influence over the terms of aesthetic and critical discourse in the early decades of the eighteenth century. What they have not examined, however, is whether there were any links—at the level either of concepts or of practices—between the performance of scientific and medical judgments and those concerning the value of literature and art. We can begin to identify such links by examining how art critics, philosophers, and medics addressed themselves to the same problem with which, as we have seen in this book, the natural philosophers of the preceding generation were so concerned. No matter how disembodied the mind might be, it depended on the unruly body and its encounters with external things for most of its ideas, and for the performance of its judgments.

Pathologies of Experience

With this thought in mind, let us return to Addison and Shaftesbury. It is true, as we have seen, that the two thinkers tended to exclude the coarser bodily pleasures from aesthetic experience. It is not often enough noted, however, that the body loomed large in virtually all their attempts to define *inappropriate* forms of experience. Consider Shaftesbury's ambiguous position on the phenomenon then known as *enthusiasm*, expressed in a letter of 1708. In the latter part of the seventeenth century, supporters of the precarious political and religious compromise established after the civil war used the term *enthusiast* to attack Nonconformist Protestants who claimed to be capable of imparting prophecies communicated to them directly by God.[16] Shaftesbury also regarded this form of enthusiasm as false, and he had no desire to encourage it. That said, he found the image of individuals spurred into action by an affective conviction of their connection to God appealing. He even argued that a genuine form of enthusiasm—a "real feeling of the Divine Presence" ultimately perceived by the mind—led to a form of mental pleasure that was in turn conducive to virtue. It was here that Shaftesbury made the connection to what we now call aesthetic experience.[17] For Shaftesbury, the experience of beauty—whether arising from the products of art or nature—was an expression of the same beauty accompanying our encounters with the divine presence. Aesthetic pleasure was both analogical and propaedeutic to genuine inspiration, leading in degrees from the enjoyment of form in material things to the rational pleasure of contemplating the mind that brought them into being. Consistent with his Platonism, moreover, Shaftesbury regarded beauty as a real phenomenon, emanating from the universal mind that gave the world order and meaning.[18]

The problem, however, was that true and false inspiration were hard to tell apart. The two states of mind both induced action through feeling rather than rational calculation, and they both led the body to assume unusual attitudes and gestures. In the "Letter Concerning Enthusiasm," Shaftesbury ventured that perhaps the only appropriate response may be to turn inward, seeking to work out whether one's own passions were in alignment with a genuine feeling of the divine presence.[19] Elsewhere in the same text, however, he ventured a criterion for distinguishing between the two forms of enthusiasm:

> I am told [. . .] that they [enthusiasts] are at this very time the subject of a choice [. . .] puppet-show at Bartholomew Fair. There, doubtless, their strange voices and involuntary agitations are admirably well acted, by the motion of wires and inspiration of pipes. For the bodies of the prophets in their state of prophecy, being not in their own power but [. . .] mere passive organs actuated by an exterior force, have nothing natural or resembling real life in any of their sounds or motions so that how awkwardly soever a puppet-show may imitate other actions, it musts needs represent this passion to the life.[20]

Shaftesbury likened the strange bodily motions displayed by enthusiast prophets to those of puppets. The puppeteers at Bartholomew Fair could accurately depict the gestures of false prophets, he suggested, because they manipulated the bodies of puppets in the same mechanical manner that the surging passions directed the bodies of enthusiasts. He thus hinted that false enthusiasm was nothing more than a pathological symptom, generated not by a genuine connection to God but instead by the distempered, mechanical motions of a sick body.[21]

Making this argument, Shaftesbury reproduced a critique of enthusiasm which, by that time, had become conventional. Regarding claims of intimacy with God as a threat to the social order, supporters of the establishment frequently portrayed enthusiasm as a form of the mental disorder known traditionally as *melancholy*—a protean psychosomatic condition with symptoms including inexplicable sadness, an obsessive focus on a single object, hallucination, and even a loss of control over the body. By the close of the seventeenth century, physicians increasingly referred to the condition by another pair of terms that had already been in use for some time—*hypochondria* if it occurred in men or *hysteria* in women. Like the term *melancholy* itself, both those names referred to the notion that the superior faculties had somehow fallen under the influence of the organs below the ribs, whether the stomach, the spleen, the genitals, or the womb.[22] In *De Anima Brutorum*, Thomas Willis thus hypothesized that religious melancholy arose when "a despair of *Eternal Salvation*" starting in the rational soul stirred up violent motions in the animal spirits. If those motions continued too long, they would eventually become fixed, continuing despite the commands of the rational soul and prompting the imagination to form false images of spiritual beings, such as angels and demons. Despite its ostensibly spiritual form, therefore, enthusiasm was just a pathological form of desire, having more in common with the lust for sex or food than a movement closer to God.[23] Shaftesbury later made this point himself in the *Miscellaneous Reflections* he included in the last volume of his *Characteristicks of Men, Manners, Opinions, Times* (1711). There, he diagnosed the false form of enthusiasm as an expression of coarse bodily impulses, insinuating that it often had the same causes as sexual desire, and likening its symptoms to "the hot and cold fits of an ague [a fever]." For Shaftesbury, therefore, the pathological body could prevent the disembodied mind from experiencing the beauty of the divine presence as it should.[24]

Shaftesbury developed his rationalist, disembodied account of aesthetic experience out of distaste for Locke's argument that the pleasures and pains of sensory experience could serve as the basis for moral life.[25] By contrast, Addison embraced Locke's description of a mind that gradually filled itself with ideas mediated by the sensory and cognitive organs. Even though Addison sought to distance it from the coarser gratifications of sex and food, he therefore still suggested in his essays on the pleasures of the imagination that aesthetic pleasure depended in some measure on the mechanisms of the body. He used Descartes's account of the passage of animal spirits through the brain, for example, to tentatively explain why pleasant memories seemed so much easier to recall than disagreeable ones. He suggested that enjoyable sensations stirred up more powerful motions in the animal spirits than those giving rise to unpleasantness, forcing wide passages through the brain that would later facilitate the rapid recollection of the sensations as memories. Proposing material causes for aesthetic pleasure, however, did not make

Addison a materialist.[26] In common with the physico-theologians, he instead wanted to show that the pleasures of the imagination were somehow encouragements to—and symptoms of—the command exercised by the soul over the body. Like Francis Hutcheson a decade or so later, he used a physico-theological argument to explain why certain objects led to a refined form of pleasure. Annexing the experience of beauty to our encounters with those objects, Addison argued, God gave humans an affective motivation to make new discoveries and to behold the works of creation with appropriate sentiments of devotion and gratitude. Addison thus concurred with the notion that genuine aesthetic pleasure could only arise when the body's affective responses were in alignment with the immaterial reasoning faculty.[27]

Like Shaftesbury, therefore, Addison raised the specter of hypochondria when he sought to explain why we sometimes experience unpleasant or inappropriate affective states. In the last of his essays on the pleasures of the imagination, for instance, he briefly acknowledged that the imagination also had an extensive capacity for pain:

> For the Imagination is as liable to Pain as Pleasure. When the Brain is hurt by any Accident, or the Mind disordered by Dreams or Sickness, the Fancy [imagination] is over-run with wild dismal Ideas, and terrified with a thousand hideous Monsters of its own framing.[28]

Without making it quite explicit, Addison invoked one of the classic symptoms of melancholy or hypochondria—the unwanted formation of terrifying images in the mind.[29] Indeed, his account of the pains of the imagination bears comparison with the description of hypochondria given by the physician Thomas Sydenham (1624–1689) in his posthumously published *Processus Integri in Morbis fere Omnibus Curandis* (1692), soon afterward translated into English as *The Compleat Method of Curing Almost all Diseases* (1694). In this work, Sydenham gave a more specific cause for the condition, arguing that it arose when some shock led the animal spirits to take on a disorderly motion. That motion in turn, he suggested, would lead sufferers to "portend the most dismal things to themselves."[30] Addison could include dreams alongside brain damage and sickness, moreover, because contemporary philosophers and physicians regarded them as images formed at random when, during sleep, the mechanisms of ideation were no longer constrained by the intellect. Like the horrifying mental images arising from a fever or a blow to the head, dreams took place when the bodily parts of the mind had too much of a role in the formation and comparison of mental images. In contrast to its pleasures, the pains of the imagination were to be regarded as pathological symptoms, unlikely to arise when the bodily parts of the mind were operating as they should.[31]

Ultimately, neither Addison nor Shaftesbury quite succeeded in separating the supposedly rational experience of beauty from the coarser impulses of their own bodies. Distancing aesthetic experience from the lower body involved making that body responsible for all the affective responses that did not seem to be in alignment with the exercise of reason. Addison and Shaftesbury both indicated, indeed, that bad taste was a form of pleasure that arose from a class of disorders in which the lower body gained undue influence over the workings of the mind. In this book, we have seen that Ray and his contemporaries made similar claims when discussing failures to enjoy the pleasures flowing from the discovery of design in nature. Identifying the failure to enjoy the experience of divine design as a symptom of some bodily disorder, those natural philosophers suggested that regimes of habituation could gradually bring the pleasures of the body into alignment with reason. For Ray and his contemporaries, the cultivation of taste was thus a literal exercise in the cultivation of the body's responses to sensory experience. I would not suggest that Addison and Shaftesbury based their own approaches to the experience of beauty on this exact model. Their interest in hypochondria, however, shows us that the attempt to disembody aesthetic experience did not necessarily entail the disembodiment of taste. On the contrary, the art critics and aesthetic theorists of the early eighteenth century engaged—to a far greater extent than has so far been acknowledged—with broadly medical or natural-philosophical practices for managing and even altering the pleasures and pains of the body.

Hypochondria and Taste

It is well known that the complex of nervous maladies increasingly described using the language of hypochondria and hysteria—which at that time were generally seen as the male and female versions of the same basic disorder—acquired new significance in the opening decades of the eighteenth century. Scholars including G. S. Rousseau, John Mullan, and G. J. Barker-Benfield have all shown that a growing interest in nervous therapeutics, motivated in part by new hypotheses about the workings of the brain and nervous system, had an important impact on the cultural politics of the day. As Barker-Benfield points out, that impact was especially pronounced in representations of the relationships between men and women. Drawing on contemporary nervous medicine and physiology, authors such as the novelist Samuel Richardson (1689–1761) suggested that the delicacy of women's bodies made them more responsive to the passions and emotions than their supposedly more rational male counterparts.[32] There has been little recognition, however, that the nervous medicine of the early eighteenth century also addressed itself squarely to the problem of taste. Physicians specializing in nervous disorders dealt directly with the question of how to alter the pleasures and pains of the body, the better to follow the judgments of reason.

Around the turn of the eighteenth century, medical writers increasingly argued that hypochondria and hysteria were caused by bad diet and poor physical regimen. Physicians specializing in nervous disorders offered a variety of explanations for the damage inflicted by these bad habits, but they consistently argued that, over time, rich food and drink corrupted the structures responsible for sensation and voluntary motion—the nerves and the muscles.[33] Although they held opposing views about the physical processes involved, for instance, the physicians John Purcell (1674–1730) and George Cheyne (1671/2–1743) both argued that foods rich in salt and liquors full of volatile spirits changed the composition of the blood. Packed with dangerous saline particles, the blood and the other fluids distilled out of it would eventually damage the nerves and their muscular coatings, making it increasingly difficult for the immaterial soul to control the body or receive accurate impressions from the external world. A sedentary life, they both suggested, made the harm even worse, depriving the nerves and muscles of the exercise needed to maintain the tone or springiness that otherwise made them reliable conduits for the transmission of impulses to and from the brain. In his book on nervous disorders in women, *A Treatise of Vapours; or, Hysterick Fits* (1702), Purcell therefore characterized hysteria as a disease of idleness and corrupt desires. Indeed, he claimed that women were more susceptible than men to nervous maladies not only because they generally led less active lives but also because they were more likely to "eat [. . .] Odd, Indigestible, suger'd [*sic*], spic'd and Salt Meats." Some thirty years later, in *The English Malady* (1733), Cheyne used similar reasoning to explain the alarming prevalence of nervous disorders among the wealthy men and women of urban England. As Cheyne saw it, the well-to-do brought hypochondria and hysteria on themselves by seeking out highly spiced foods that harmed the body and corrupted the appetites and by spending too much of their leisure time sitting down.[34]

Although physicians specializing in nervous disorders prescribed pharmaceutical remedies and other medical interventions for the worst hypochondriac symptoms, whether extreme anxiety, delirium, or convulsions, they usually urged that better habits of diet and exercise were the only way to obtain a durable cure. This course of action usually involved cutting down on red meat and strong spirits in favor of plainer fare, while taking moderate forms of exercise such as walking and riding, or perhaps mounting a spring-loaded chair designed to mimic the effects of being carried by a horse.[35] Joanna Picciotto has shown that the regimens prescribed by Cheyne and Purcell, among others, extended the Royal Society's program for reversing the effects of Adam's fall from grace on the human faculties to a broader public. Much as Hooke saw the microscope as an instrument of regeneration, they depicted diet and exercise as means of making the body into a more effective vehicle for the exercise of reason. Rather than simply reacting to its inaccurate perceptions and irrational urges, the improved body would transmit clear information to the mind and obediently implement judgments made by the rational soul.[36] As Anita Guerrini has also pointed out, Cheyne made spiritual regeneration central to his portrayal of the effects of diet and exercise. Insisting that the cause of virtually all diseases was sinful consumption arising from corrupt bodily desires, he saw the pursuit of dietary regimen as a means of making the body a fitter instrument for virtue. In *The Natural Method of Cureing the Diseases of the Body, and the Disorders of the Mind Depending on the Body* (1742), moreover, he asserted that "that *Pains*, *Suffering*, and *Diseases*, are necessary in the *Oeconomy* of *Providence*, to make Men virtuous, in order to become afterwards happy [. . .]." God had, Cheyne argued, instituted the painful symptoms of disease as instruments of affective persuasion. Perhaps incapable of using their reason to distinguish between the false pleasures of luxury and the more durable pleasures of virtue, sufferers might still be persuaded by the torments of nervous disorders to undertake the seemingly unpleasant regimen required to make themselves healthy, first in body and then in spirit.[37]

Building on Picciotto's demonstration, I would suggest two further ways in which the therapies prescribed for nervous disorders in the early eighteenth century addressed the aesthetic dilemmas so important both to the empirical sciences of the preceding decades and to contemporaneous aesthetics and art criticism. First, like the physico-theologians, physicians presented regimen as a practice that worked by modifying the body's affective responses to sensory experience, gradually enabling it to enjoy what would do it good and to dislike the harmful things that had before given it so much pleasure. In his *Medicina Gymnastica* (1705), for example, the physician Francis Fuller (1670–1706) identified riding and other forms of exercise as the most effective therapy for a range of disorders that included hypochondria. Acknowledging that this course of physical exertion might not much appeal to those accustomed to sedentary luxuries, he offered reassurance that repeated practice would soon make what had once seemed unpleasant into a source of enjoyment:

> so that an Active Life, when a Man has laid aside his timourous Prejudices, and is let into the tast of it, will be found not only to have its Advantages, but its Charms too; and he who indulges himself long in it, will think it not a Paradox, that there should be an Active Luxury, which may exceed all the Passive Enjoyments of Sloth and Indolence.[38]

Cheyne made a similar case in the closing pages of his *Natural Method*, offering the last in an endless series of encouragements to take up his diet of vegetables, milk, and seeds even though it would be unlikely to taste good at first. "Although the method be *slow*, and somewhat *self-denying*," he wrote, "yet *Custom* will make it still easier, and the *Health* and *Spirits* arising from it will in time make it *pleasant*." The purpose of regimen was not simply to eliminate the causes of disease. Its broader aim was to transform the body's noncognitive or irrational responses to external objects, leading it to enjoy forms of consumption and embodiment coinciding with—but not directly controlled by—the judgments of the rational soul.[39]

Second, like the natural philosophers and aesthetic theorists we have considered in the body of this book, medics specializing in nervous disorders generally promoted a brand of mind–body dualism that broke down at the level of practice. Despite holding that the soul was an immaterial entity, they made it dependent on the state of the body for the exercise of all its functions. As Picciotto has noted, Purcell thus argued that nervous diseases could make the exercise of reason impossible, depriving the soul of the clear and distinct ideas it needed to make accurate judgments.[40] In addition to Purcell, meanwhile, Fuller and Cheyne both pointed out that the same disorders could prevent the soul from exercising control over voluntary motion, leaving it hostage to the body's irrational impulses.[41] It was perhaps Cheyne, however, who gave this ambiguous dualism its most telling expression. In both the *English Malady* and *Natural Method*, he did not—as we might expect—suggest that the ideal body for the exercise of reason was either completely healthy or entirely under the command of the soul. On the contrary, he flattered his wealthy readers by suggesting that the state of nervous delicacy engendered by their luxurious lives could have beneficial effects if prevented from becoming too severe. The upper classes, he suggested, had "more delicate and *elastic Organs* of *Thinking* and *Sensibility*" than those people whose brains and nerves were worn down by endless manual labor. The wealthy, Cheyne continued, had access a world of rational pleasures far more refined, and more conducive to virtue, than the sensual pleasures to which working people were confined. The putatively rational disposition to virtue was therefore almost as embodied as the supposedly irrational appetite for the pleasures of food and sex. It was not enough to have a soul that decided on and willed acts of reason and goodness. The capacity to enjoy, and thus pursue, virtue arose from a bodily constitution that could lead to sinful overindulgence if not properly regulated.[42]

The nervous medics of the early eighteenth century therefore held that the soul could not function as it should without the cooperation of the body and the affective states to which that body was liable. Like the natural philosophers of the preceding decades, they responded to this predicament by arguing that the exercise of rational thought depended to a large extent on undertaking forms of habituation that would gradually make the organs of sensation and cognition enjoy what would do them good. In other words, they also addressed the problem of getting people to agree about the apparently subjective feelings arising from the body's encounters with the external world. To an even greater extent than Ray and his contemporaries, they responded to this problem with a highly literal and thus embodied account of taste.

At this stage, the conclusions we may draw from tracing the same problem across three fields—natural philosophy, aesthetics, and medicine—are of course speculative. However, I believe that this shared concern with the pleasures and pains of sensory experience opens a path for future inquiry. It suggests that during the early eighteenth century, the two meanings of taste highlighted by Agamben—the literal sense of gustation and the metaphorical sense of appreciating beauty—may in fact have been more closely connected that has so far been understood. Most accounts of aesthetics in the early eighteenth century, as we have seen, focus on processes thought to take place in the mind, at some distance from the coarser desires and impulses of the body. Yet a rather more literal process of bodily habituation may also have been involved, requiring individuals to change rather than simply overcome their feelings of pleasure and pain. Even those who sought to distance the pleasures of aesthetic experience from those of the lower body found profitable ways to mobilize practices for managing the pleasures and pains of sensory experience. Proving this point would be the work of another book, but I would like to end with giving one example of how we might use the approach outlined here to reconsider the links between aesthetics, the empirical sciences, and medicine in the eighteenth century.

Regimen and the Science of the Connoisseur

It is well known that the portraitist Jonathan Richardson (1667–1745) made extensive use of Locke's empiricism in his theory of art, given its clearest expression in his influential *Two Discourses* on connoisseurship (1719). Publishing them together in the same volume, he called the first *An Essay on the Whole Art of Criticism as it Relates to Painting*, and the second *An Argument in behalf of the Science of a Connoisseur.*[43] As the title of the second discourse indicates, Richardson's aim was to present the judgment of works of art as an empirical science, capable of producing the same level of certainty then associated with natural philosophy.[44] Carol Gibson-Wood has shown that Richardson made his case by taking Locke's account of the unpleasant work of judgment—the one we encountered in chapter 4—as the model for judging the beauty and originality of paintings and drawings. To become connoisseurs, individuals needed to begin by stocking their minds with ideas derived from studying works of art. Rather than taking immediate pleasure from those ideas, however, they needed to train the mind to make fine distinctions between things that might at first seem very similar.[45] To see how Richardson thought this procedure would work out in practice, consider his approach to attribution. It was first necessary, he suggested, to fill the mind with images of works known to be by the artist in question. Only by gradually building up such a mental picture, Richardson argued, would it then be possible to judge—sometimes focusing on minute details such as the handling of fingers and toes—whether the work was by that artist or another. Rather than taking pleasure in the most obvious similarities, a true connoisseur would carefully compare the work of art in question with her or his stock of mental images taken from authentic paintings and drawings.[46]

We are thus left with the impression that Richardson saw connoisseurship as a dispassionate empirical science, dependent on the mental discipline of making minute distinctions between ideas rather than engaging with the pleasures and pains of the body. That impression is to some extent contradicted, however, by Richardson's persistent concern that the irrational appetites of the body could disrupt such fine judgments. He addressed this possibility throughout the *Two Discourses* but made it his focus in a lengthy section dealing with the pleasures of connoisseurship at the close of the second discourse. As Gibson-Wood has shown, Richardson based many of the positions he would take in this section on an unpublished physico-theological poem, entitled "Hymn to God," that he composed during the winter of 1711–1712.[47] In that earlier text, Richardson asserted that self-love—the desire for happiness—motivated all human acts, including acts of virtue. The difficulty, however, was that the rational soul had little authority over that most powerful of impulses:

> Weak, Impotent the whole in Chains secur'd,
> Subjected to Involuntary Acts;
> Or to Ideas, whether from within
> Uprais'd Spontaneous, (seeming so at Least)
> Or else injected by External things
> Seen, Tasted, Swallow'd, Heard, or Felt, or Smelt,
> Producing Doubts, Enquirys, Debates,
> Determinations, Passions manifold,
> Desires, Aversions, Wills, & Lastly Acts.
> [. . .] so man
> (As other Animals) by Fate endued
> With Selfe Love from his Being inseparable
> A Part of him; whatever Principle
> Subordinately under Wild Passion, or
> Reason Sedate, Selfe Love is still in view.[48]

In these indigestible verses, Richardson began by asserting that many seemingly voluntary acts are nothing more than involuntary responses generated by the body from its encounters with external things. He continued by suggesting that it might be hard to tell those two categories of action apart. Whether prompted by reason or provoked by the body's passionate impulses, every single human action sprang from some form of self-love.[49]

In the *Two Discourses*, Richardson made it clear that he saw the mind as an immaterial entity, identifying the pleasures to be had from contemplating works of art as rational pleasures of the soul. He nevertheless made the difficulty of distinguishing between those pleasures and the gratifications of the body central to his remarks on the practice of connoisseurship. Indeed, he opened the first discourse by translating his earlier musings on the causes of human action into the language of contemporary nervous medicine, remarking that the mind was frequently led astray by impressions made on the senses and alterations taking place in the "Fluids and Solids" of the body.[50] In the section of the second discourse dealing with the pleasures of connoisseurship, meanwhile, he repeated his suggestion that individuals might find themselves mistaken about their own feelings of pleasure. "The Desire of Happiness," he wrote, "is the Spring that puts us all in Motion; We receive it together with the Breath of Life [. . .] This is the End in which we All agree, tho' as to the Way there is infinite Variety, and Error."[51] In the following pages and elsewhere, Richardson therefore sought to show that the pursuit of virtue would lead to the most satisfying and durable pleasures of all. At the same time, however, he needed to explain why some people mistakenly chose the misleading delights of the lower body, or even failed entirely to experience gratification when they should. Like Addison, he found an answer in the pathologies of consumption, arguing that those who overindulged would eventually suffer from painful maladies to which the virtuous were invulnerable. He also discussed some of the physical causes that lead people to feel differently about the things around them, pointing out the influence of alcohol and tobacco over perception, and remarking that maladies such as hypochondria and gout sometimes made it impossible to enjoy things that had previously brought pleasure.[52]

Like the medics specializing in nervous disorders, Richardson asserted that the way to bring the pleasures of mind and body into alignment with right reason was through regimen. The mental labor of connoisseurship was, he suggested, such a regimen. It was a means not only for learning how to distinguish beauty from ugliness, or originality from imitation, but also for gaining the capacity to take pleasure in things that might not at first seem enjoyable. For Richardson, moreover, there were close parallels between the acquisition of good taste in art and the more obviously embodied form of taste at stake in eating and drinking. The connoisseurship of paintings and drawings, he suggested, was part of a broader exercise in temperance that began with food and drink. Echoing the nervous medics, he thus argued that dietary restraint would do more than heal the body. It would transform the palate: "By Temperance, and Sobriety a Common Meal is a Feast for an Epicure. True Rational Appetite turns Water into Wine, and every Glass is *Tokay*."[53] Further on, he applied the same lesson to art appreciation, pointing out that its chief benefit was gaining the capacity experience pleasures that lay beyond the grasp of those who had not undergone the same discipline. Richardson explained that "A *Connoisseur* has this farther Advantage, He not only sees Beauties in Pictures, and Drawings, which to Common Eyes are Invisible; He Learns by these to see such in Nature [. . .] and from whence he has a Pleasure which otherwise he could never have had."[54] For Richardson, filling the mind with images of paintings and drawings and learning to discriminate minutely between them promised the same benefits as sticking to plain food and drink. Whether pursued through diet or the study of art, regimen would remedy the body's propensity to irrational and pathological forms of pleasure, while enabling it to experience the more refined gratifications accompanying the exercise of reason and concomitant pursuit of virtue.

Richardson's science of connoisseurship more closely resembles the aesthetic science we have encountered in the previous chapters than the dispassionate exercise in the production of knowledge described by Gibson-Wood. Indeed, we might very well read Richardson's aesthetic regimen as a response to Locke's suggestion that the only way to make pleasure the foundation for good conduct was by making people "correct their palates, and give a relish to what either has, or they suppose has none." Richardson's model of connoisseurship thus belonged in some measure to the aesthetic world inhabited not only by philosophers such as Boyle, Hooke, and Ray, but also by physicians including Purcell, Fuller, and Cheyne. No matter how disembodied the mind might be, the exercise of reason was tightly bound up with the pleasures and pains of the body. As we have seen, the design argument gave this account of experience an ontological foundation, using God's beneficent involvement with the world to underpin what often seemed like a precarious alignment of the discovery of the truth, the pursuit of virtue, and the experience of pleasure. For Richardson, as for Addison and Shaftesbury, divine design explained why certain kinds of objects generally induced the experience of beauty and why aesthetic judgments could be found true or false. In the final analysis, we might say that the design argument made it possible for philosophers of the seventeenth and eighteenth centuries to ground both their theories of aesthetic experience and their sciences of nature on the apparently subjective basis of taste. In a world where some judgments of taste reflected an order emanating from the wisdom and goodness of God, the experience of aesthetic pleasure could be considered in some sense objective. It was, in other words, a place where an aesthetic science was possible.

Aesthetics and Intersubjectivity

Ultimately, this book makes the case for new ways of thinking about the early modern arts and sciences. Rather than examining the links between natural philosophy and individual domains such as literature or the visual arts, I have used the category of aesthetic experience to bring a far wider range of disciplines into focus. Indeed, I have suggested that we can only grasp how early modern thinkers reckoned with the affective dimensions of experience by bringing together the whole range of disciplines and practices then at stake, from rhetoric and poetics to theology and metaphysics. Using this approach, we have seen that a group of philosophers associated with the Royal Society of London took a keen interest in the affective dimensions of experience. Against the view that they sought a recognizably modern form of objectivity, John Ray and his contemporaries saw the discovery of the truth as an exercise in the cultivation of taste. They wanted to harness the pleasures and pains of the body to their philosophical project, and they identified the practices of natural philosophy as means to alter those feelings for the better. I have suggested, moreover, that such a model of empiricism may have had a far more important role in the development of aesthetics and art criticism than has been recognized so far. It was not just later materialists such as Edmund Burke and Denis Diderot who engaged with medical explanations and practices concerning the effects of sensory experience on the body. Earlier thinkers such as Joseph Addison, Francis Hutcheson, the Third Earl of Shaftesbury, and Jonathan Richardson are usually thought to have anticipated Immanuel Kant's disembodied account of aesthetic experience. In different ways, however, they all turned to the resources of nervous medicine in their discussions of aesthetic taste.

More important, we have seen that the aesthetic is a powerful conceptual tool for reckoning with one of the fundamental questions in the history and philosophy of science. Among historians of science, it is now common to argue that the history of scientific objectivity should be understood as the history of intersubjectivity. Lorraine Daston and Peter Galison have shown, for instance, that the image of scientific objectivity has changed enormously over the past five hundred years, as natural philosophers identified it with new intellectual practices and techniques for recording and disseminating information.[55] So far, however, there has been little recognition that we might enhance our understanding of intersubjectivity by turning to the intellectual resources of aesthetics and taste. Yet the production of intersubjective agreement is one of the central questions addressed by theorists of both aesthetics and gustation alike. Whether thinking about the feelings caused by food and medicine, or those arising from the contemplation of art, philosophers and practitioners have used taste to discuss the problem of getting people to agree about the apparently subjective dimensions of individual experience. By following the historical actors and collapsing the customary distinction between taste and gustation, we may therefore imagine a new history of intersubjectivity. Such a history would examine the kinds of feelings scientific practitioners expected from the study of nature, and what strategies they used to obtain agreement about them. For Ray and his contemporaries, at any rate, the community of knowledge was a community of feeling. To grasp the truth was to learn how to enjoy it.

Acknowledgments

Writing this book, I have been lucky to enjoy far more support and encouragement than I could ever have expected. I have been continually surprised by the willingness of funding bodies, both private and public, to make it possible for me to undertake my work. I am thus delighted to acknowledge the indispensable material support given to me by the Arts and Humanities Research Council, the Institute of Historical Research, the Max-Planck Institute for the History of Science, the Huntington Library, the California Institute of Technology, and the Leverhulme Trust. In turn, I would like to thank both the taxpayers and private benefactors who have made it possible for those organizations to act with such generosity.

This project would also have been impossible without the resources made accessible to me by wonderful libraries and their staff. For making this book possible, I am very grateful to all those who work in the Bodleian Library, the British Library, Cambridge University Library, the Huntington Library, the New York Public Library, and the libraries of St. Catharine's College, St. John's College, and Trinity College, Cambridge.

Working on *Aesthetic Science*, I have benefited immensely from the insight, encouragement, and friendship of my teachers and mentors. I owe a special thanks to Anne Goldgar, who first nourished my interest in early modern thought when I was a student at King's College London, and Simon Schaffer, who supervised my studies at Cambridge. In addition, I would like to extend special thanks to Ludmilla Jordanova, Stephen Clucas, and John Brewer for all they have done on my behalf.

Over the years, I have been fortunate to meet some remarkably generous kindred spirits, intellectual allies, and interlocutors. I am especially thankful to Joanna Picciotto for taking me under her wing, inspiring me with her encouragement, enthusiasm, and profound learning. I would also like to thank Frédérique Aït-Touati, Ben Breen, Michael Bycroft, Surekha Davies, Sarah Easterby-Smith, Elizabeth Eger, Stephen Gaukroger, Ardeta Gjikola, Florence Grant, Felicity Henderson, Matthew Hunter, Michael Hunter, Sarah Tindal Kareem, Rhodri Lewis, Lan A. Li, Daniel Margocsy, Alex Marr, Brian W. Ogilvie, William Poole, Carmel Raz, Anna Marie Roos, Courtney Weiss Smith, William Tullett, and Claude Willan for their generosity and thoughtfulness.

I first began to conceive of this book when I was in the Department of History and Philosophy of Science at the University of Cambridge. For making my time in that extraordinary department so rewarding, I would like to thank Alexandra Bacopoulos-Viau, Geoffrey Belknap, Mirjam Brusius, Anita Herle, Tamara Hug, Nicholas Jardine, Sachiko Kusukawa, Lavinia Maddaluno, Iris Montero, Deirdre Moore, Jenny Rampling, Anne Secord, Richard Serjeantson, Katie Taylor, Nick Whitfield, and Kelly Whitmer.

Like many scholars today, I completed this book while working through a long series of temporary teaching positions and short-term fellowships. During my time at the Max Planck Institute for the History of Science in Berlin, I gained much from discussions with Etienne Benson, Lorraine Daston, Susan Naquin, Skuli Sigurdsson, and Fernando Vidal. Special thanks also to Ryan Kashanipour and Marta Jordi Taltavull for their warm friendship and crucial support. A few years later, I enjoyed a transformative nine months as a research fellow at the California Institute of Technology and the Huntington Library. My time at those two institutions had a decisive impact on my work, giving me both the time and the intellectual climate to discover what form this book should take. For making that experience possible, I am thankful especially to John Brewer, Juan Gomez, Steve Hindle, Theresa Kelley, and Joanna Picciotto. I would also like to extend my thanks to Daniela Bleichmar, Timothy Costelloe, Mordechai Feingold, Kristine Haugen, Phil Hofer, Lisa Jardine, Jeanette Kohl, Jonathan Lamb, Lyle Massey, Natania Meeker, Lindsay O'Neill, Lucy Razzall, James Simpson, Joanna Stalnaker, Michael Yonan, and Rosie Young.

During my year as a lecturer at Somerville College, University of Oxford, I had the opportunity to teach alongside some wonderful colleagues. I am immensely grateful to Joanna Innes, Nicholas Davidson, and Benjamin Thompson for making me a better teacher and for welcoming me so warmly to the Oxford intellectual life. I owe the same debt of gratitude to the colleagues and friends I made during the time I spent teaching in the Department of History at New York University, especially Karl Appuhn, Myles Jackson, Anne O'Donnell, Guy Ortolano, and Benjamin Schmitt. Finally, I worked for University College in three different capacities while working on this book—twice as a teaching fellow and once as a research fellow. For their support during my time in London, I would like to thank Chiara Ambrosio, Selina and John Eger, Mechthild Fend, Claire Morley, Julietta Steinhauer, and Simon Werrett.

I thank my students, whose kindness and intelligence kept me going through difficult times. I am especially grateful to Joe Shalam for his enthusiasm and messages of encouragement. I am also very grateful to all the people associated with the University of Chicago Press who have made possible the publication of *Aesthetic Science*. I am especially grateful to Karen Darling for supporting my work and to Lori Meek Schuldt for her insightful copyediting. I am, moreover, delighted to acknowledge the anonymous reviewers, whose perceptive comments enabled me to improve this book.

I could never have completed *Aesthetic Science* without the support of my friends and family, past and present. For their friendship over the years, I am grateful to Sophie Brockmann, Edward Cohen, Caroline Fries, Matthew and Jennifer Heap, David Lucas, Hannah Malone, Kathryn Santner, Jake Snyder, and Daniel Street. For welcoming me to their family, meanwhile, I thank my kind parents-in-law, Marie-France and Louis-Bertrand Bishop. I would also like to thank my aunt, Susan Williams, who has patiently waited for me to finish this book, expressing endless encouragement along the way. My mother, Catherine W. Morley, first encouraged me to pursue history, philosophy, and art. I will be forever grateful to her for introducing me to that world.

Finally, thank you, Cécile, for helping me so much, and for choosing to live your life with me. I would not have wanted to write this book without you.

Parts of chapter 2 were published in a previous form as "Robert Boyle and the Representation of Imperceptible Entities," *British Journal for the History of Science* 54, no. 1 (2018): 17–40. Smaller portions of chapters 5 and 6, meanwhile, appeared in earlier versions as "'Vividness' in English Natural History and Anatomy," *Notes and Records of the Royal Society* 66, no. 4 (2012): 341–56, and "The Work of Verbal Picturing for John Ray and Some of His Contemporaries," *Intellectual History Review* 20, no. 1 (2010): 165–79. I am grateful to the publishers of those journals for permission to publish revised versions of these works here.

Notes

Introduction

1. This translation is by Matthew L. Jones, given in Matthew L. Jones, *The Good Life in the Scientific Revolution: Descartes, Pascal, Leibniz, and the Cultivation of Virtue* (Chicago: University of Chicago Press, 2006), 215. Jones was translating the original found in Gottfried Leibniz, *Sämtliche Schriften und Briefe*, 6th ser., vol. 3 (Darmstadt: Deutsche Akademie der Wissenschaften, 1923), 433–34.

2. Joseph Glanvill, *Plus Ultra; or, The Progress and Advancement of Knowledge since the days of Aristotle* (London, 1668), 52.

3. Robert Hooke, "A General Scheme, or Idea of the Present State of Natural Philosophy, and how its Defects may be Remedied by a Methodical Proceeding in the making Experiments and collecting Observations. Whereby to Compile a Natural History, as the Solid Basis for the Superstructure of True Philosophy," in Robert Hooke, *The Posthumous Works of Robert Hooke*, ed. Richard Waller (London, 1705), 8.

4. John Locke, *An Essay Concerning Human Understanding*, ed. Peter H. Nidditch (Oxford: Oxford University Press, 1975), bk. 2, ch. 21, pp. 251–82.

5. Peter Walmsley, *Locke's Essay and the Rhetoric of Science* (Lewisburg, PA: Bucknell University Press, 2003), 109–11.

6. See for instance Locke, *Essay*, bk. 2, ch. 21, sec. 69, pp. 280–81.

7. For a long time, historians and philosophers of science argued that the mathematization of experience, most salient in the developments in physics and astronomy from Galileo to Newton, was what led to the emergence of the modern empirical sciences. See E. A. Burtt, *The Metaphysical Foundations of Modern Physical Science* (London: Routledge and Kegan Paul, 1924); and Alexandre Koyré, *Etudes Galiléennes* (Paris: Hermann, 1939). For more up-to-date treatments, see Peter Dear, *Discipline and Experience: The Mathematical Way in the Scientific Revolution* (Chicago: University of Chicago Press, 1995); and especially Stephen Gaukroger, *The Emergence of a Scientific Culture: Science and the Shaping of Modernity, 1210–1685* (Oxford: Oxford University Press, 2006). The notion that the emergence of the modern empirical sciences depended on disciplining the subjectivities of individual experience has also exercised considerable influence in other fields in the history of science. In her important study of Dutch visual culture in the

seventeenth century, Svetlana Alpers argues that both science and the descriptive style of art epitomized by the still life genre were symptomatic of a desire to picture the world with the same accuracy as optical instruments such as the camera obscura. See Svetlana Alpers, *The Art of Describing: Dutch Art in the Seventeenth Century* (Chicago: University of Chicago Press, 1983). In his history of botany in the Renaissance, meanwhile, Brian W. Ogilvie argues that botany came to be evacuated of moral and affective significance with the emergence of large-scale systems of classification in the seventeenth and eighteenth centuries. See Brian W. Ogilvie, *The Science of Describing: Natural History in Renaissance Europe* (Chicago: University of Chicago Press, 2006).

8. Lorraine Daston, "On Scientific Observation," *Isis* 99, no. 1 (2008): 97–110; Jessica Riskin, *Science in the Age of Sensibility: The Sentimental Empiricists of the French Enlightenment* (Chicago: University of Chicago Press, 2002); Charles T. Wolfe and Ofer Gal, "Embodied Empiricism," in *The Body as Object and Instrument of Knowledge*, ed. Charles T. Wolfe and Ofer Gal (Dordrecht, Neth.: Springer, 2010), 1–5.

9. Immanuel Kant, *Critique of the Power of Judgment*, ed. Paul Guyer, trans. Paul Guyer and Eric Matthews (Cambridge: Cambridge University Press, 2000), 24–28, 102–4, 185–95. As an example, consider Kant's reason for excluding scientific accomplishment from his account of genius. For Kant, a genius was an artist who did not possess the ability to explain the process by which her or his ideas came about. He thus excluded great natural philosophers such as Isaac Newton from his definition of genius, arguing that their ability to explain the reasoning behind their discoveries betokened a form of cognitive activity unlike that involved in either the production or the judgment of beauty. See Kant, *Critique of Judgment*, 187; and Bryan Hall, "Kant on Newton, Genius, and Scientific Discovery," *Intellectual History Review* 24, no. 4 (2014): 539–42.

10. Gregory J. Seigworth and Melissa Gregg, "An Inventory of Shimmers," in *The Affect Theory Reader*, ed. Melissa Gregg and Gregory J. Seigworth (Durham, NC: Duke University Press, 2010), 1–25, at 1.

11. Aristotle, *Art of Rhetoric*, trans. J. H. Freese (Cambridge, MA: Harvard University Press, 1926), 173.

12. In English translations of Spinoza's *Ethics*, the term *affectus* is usually translated, to misleading effect, with the word *emotion*. See, for instance, Baruch Spinoza, *Ethics*, ed. and trans. G. H. R. Parkinson (Oxford: Oxford University Press, 2000), 223. Cf. "Affectus, qui animi Pathema dicitur, est confusa idea, quâ Mens majorem, vel minorem sui Corporis, vel alicujus ejus partis existendi vim, quàm antea, affirmat, & quâ data ipsa Mens ad hoc potiùs, quàm ad illud cogitandum determinatur," in Baruch Spinoza, *Opera Posthuma* (Amsterdam, 1677), 159. As we shall see further on, thinkers of the seventeenth century generally understood the feelings that we regard as emotions to be motions operative upon or taking place within the body. For the most part, medics and natural philosophers referred to those feelings using the language of the "passions"—for instance, regarding anger not simply as a mental state but also as a movement of bodily fluids toward the bodily extremities. Although Spinoza distinguished between affects and passions, he nevertheless sought, using the term *affect*, to describe motions that resulted in action rather than the mental states that we would recognize as emotions. See Drew Daniel, "Self-Killing and the Matter of Affect in Bacon and Spinoza," in *Affect Theory and Early Modern Texts: Politics, Ecologies, and Form*, ed. Amanda Bailey and Mario Digagni (New York: Palgrave Macmillan, 2017), 96.

13. Kant, *Critique of Judgment*, 96–101.

14. Alexander Gottlieb Baumgarten, *Aesthetica* (Frankfurt an der Oder, 1750), 1, "AESTHETICA [. . .] est scientia cognitionis sensitivae." Cf. Baumgarten, *Meditationes Philosophicae de*

Nonullis ad Poema Pertinentibus (Magdeburg, Ger., 1735), 39. See also Paul Guyer, *A History of Modern Aesthetics*, 3 vols. (Cambridge: Cambridge University Press, 2015), 1:327–29.

15. Francis Hutcheson, *An Inquiry into the Original of Our Ideas of Beauty and Virtue* (London, 1725), 27. See Peter Kivy, *The Seventh Sense: Francis Hutcheson and Eighteenth-Century British Aesthetics*, 2nd ed. (Oxford: Oxford University Press, 2003), 99–101.

16. See especially Alpers, *Art of Describing*, 72–74; Steven Shapin and Simon Schaffer, *Leviathan and the Air-Pump: Hobbes, Boyle, and the Experimental Life* (Princeton, NJ: Princeton University Press, 1985), 18, 321–22; Peter Harrison, *The Fall of Man and the Foundations of Science* (Oxford: Oxford University Press, 2007), 199–201; and Matthew C. Hunter, *Wicked Intelligence: Visual Art and the Science of Experiment in Restoration London* (Chicago: University of Chicago Press, 2013).

17. Meghan C. Doherty, "Discovering the 'True Form': Hooke's *Micrographia* and the Visual Vocabulary of Engraved Portraits," *Notes and Records of the Royal Society* 66, no. 3 (2012): 228. For a perspective that draws on Hooke's use of contemporary rhetoric, see John T. Harwood, "Rhetoric and Graphics in *Micrographia*," in *Robert Hooke: New Studies*, ed. Michael Hunter and Simon Schaffer (Woodbridge, UK: Boydell Press, 1989), 119–47.

18. Robert Hooke, *Micrographia; or, Some Physiological Descriptions of Minute Bodies* (London, 1665). He describes objects as "beautiful" (or using another word including the *beaut* prefix) on 2, 80, 85, 90, 131, 140–41, 152, 160, 163, 169, 172, 174, 182, 184, 195, and 210. The word *curious* comes up throughout the text, and Hooke uses the expression "exceeding pleasant" on 49 and 152.

19. Hooke, *Micrographia*, 2.

20. Hooke, *Micrographia*, preface, sig. [a1r].

21. In the seventeenth and eighteenth centuries, scientific proofs for the existence and activity of God—including the design argument—were generally pursued in the form of treatises and sermons on natural theology, a genre that will be considered in detail in chapter 1. On the differences between the design argument made by Robert Boyle and the one made by Isaac Newton and his followers, see John Hedley Brooke, *Science and Religion: Some Historical Perspectives* (Cambridge: Cambridge University Press, 1991), 176–89; and Scott Mandelbrote, "Early Modern Natural Theologies," in *The Oxford Handbook of Natural Theology*, ed. John Hedley Brooke, Russell Re Manning, and Fraser Watts (Oxford: Oxford University Press, 2013), 87–91. Although the distinction between these two varieties of the design argument was significant in the seventeenth century (indeed, Boyle explicitly rejected the possibility of using the heavens to argue for design in nature), those writing after Newton tended to mix them together or mobilize them alongside each other.

22. William Poole, *The World Makers: Scientists of the Restoration and the Search for the Origins of the Earth* (Oxford: Peter Lang, 2010); Paolo Rossi, *The Dark Abyss of Time*, trans. Lydia G. Cochrane (Chicago: University of Chicago Press, 1984), 33–75; Rhodri Lewis, *Language, Mind and Nature: Artificial Languages in England from Bacon to Locke* (Cambridge: Cambridge University Press, 2007); William Poole, "The Divine and the Grammarian: Theological Disputes in the 17th-Century Universal Language Movement," *Historiographica Linguistica* 30, no. 3 (2008): 273–300.

23. This point will be discussed at length in chapter 1.

24. Thomas Sprat, *The History of the Royal-Society of London for the Improving of Natural Knowledge* (London, 1667), 52–54, 358–78. See also Shapin and Schaffer, *Leviathan and the Air-Pump*, 283–331; and Brian Vickers, "The Royal Society and English Prose Style: A Reassessment,"

in *Rhetoric and the Pursuit of Truth: Language Change in the Seventeenth and Eighteenth Centuries*, ed. Brian Vickers and Nancy S. Struever (Los Angeles: Clark Memorial Library, University of California, 1985).

25. William Poole, "Francis Lodwick's Creation: Theology and Natural Philosophy in the Early Royal Society," *Journal of the History of Ideas* 66, no. 2 (2005): 245–63; Rhodri Lewis, *William Petty on the Order of Nature: An Unpublished Manuscript Treatise* (Tempe: Arizona Center for Medieval and Renaissance Studies, 2012).

26. Charles Raven, *John Ray, Naturalist: His Life and Works* (Cambridge: Cambridge University Press, 1942), 59–61.

27. Michael Hunter, "Early Problems in Professionalizing Scientific Research: Nehemiah Grew (1641–1712) and the Royal Society, with an Unpublished Letter to Henry Oldenburg," *Notes and Records of the Royal Society of London* 36, no. 2 (1982): 190.

28. On Willis's royalism and High Church Anglicanism, see Louis Caron, "Thomas Willis, the Restoration, and the First Works of Neurology," *Medical History* 59, no. 4 (2015): 525–53. Willis's hypotheses about the workings of the mind will be considered in chapters 2 and 5.

29. I first became aware of this position from Ethan Shagan's study of the English Reformation, wherein he criticizes approaches that measure the impact of the Reformation by attempting to assess the extent to which individuals believed—or did not believe—in the new doctrines to which they were exposed. He argues instead that individuals participated in the Reformation, whether stripping monasteries or defending the old faith, without necessarily holding beliefs consistent with the theological positions to which the opposing sides ostensibly adhered. See Ethan Shagan, *Popular Politics and the English Reformation* (Cambridge: Cambridge University Press, 2003), 1–23. For an example more directly relevant to the histories of philosophy and science, see Jonathan Sheehan, "Thomas Hobbes, D.D.: Theology, Orthodoxy, and History," *Journal of Modern History* 88, no. 2 (2016): 249–74.

30. M. Jones, *Good Life in the Scientific Revolution*.

31. Sorana Corneanu, *Regimens of the Mind: Boyle, Locke, and the Early Modern Cultura Animi Tradition* (Chicago: University of Chicago Press, 2011); Joanna Picciotto, *Labors of Innocence in Early Modern England* (Cambridge, MA: Harvard University Press, 2010).

32. Jeffrey Chipps Smith, ed., *Visual Acuity and the Arts of Communication in Early Modern Germany* (Aldershot, UK: Ashgate, 2014), 1–19. I first became aware of this book through a review article: Alex Marr, "Knowing Images," *Renaissance Quarterly* 69, no. 3 (2016): 1009. Much of the best work on the place of religion in the visual culture of early modern natural philosophy deals with Jesuit science. See, for instance, Mark A. Waddell, *Jesuit Science and the End of Nature's Secrets* (London: Routledge, 2015). On the interrelationships between changing religious ideas and changing accounts of vision in the early modern period, see Stuart Clark, *Vanities of the Eye: Vision in Early Modern Culture* (Oxford: Oxford University Press, 2007), esp. ch. 5 and 6, pp. 161–235.

33. For Wren on architecture and natural philosophy, see J. A. Bennett, *The Mathematical Science of Christopher Wren* (Cambridge: Cambridge University Press, 1982), 87–124. On Hooke, architecture, and natural philosophy, see Matthew Hunter, *Wicked Intelligence*, 188–221. See also Matthew Walker, *Architects and Intellectual Culture in Post-Restoration England* (Oxford: Oxford University Press, 2017); and Hentie Louw, "The 'Mechanick Artist' in Late Seventeenth Century English and French Architecture: The Work of Robert Hooke, Christopher Wren and Claude Perrault Compared as Products of an Interactive Science/Architecture Relationship," in *Robert*

Hooke: Tercentennial Studies, ed. Michael Cooper and Michael Hunter (Aldershot, UK: Ashgate, 2006).

34. On Aubrey's antiquarianism, see Kelsey Jackson Williams, *The Antiquary: John Aubrey's Historical Scholarship* (Oxford: Oxford University Press, 2016). For a more general discussion, see Craig Ashley Hanson, *The English Virtuoso: Art, Medicine, and Antiquarianism in the Age of Empiricism* (Chicago: University of Chicago Press, 2009), 58–92. The best overview of the society's engagement with what we would call archaeology is still Michael Hunter, "The Royal Society and the Origins of British Archaeology: I," *Antiquity* 45, no. 178 (1971): 113–21; and "The Royal Society and the Origins of British Archaeology: II," *Antiquity* 45, no. 179 (1971): 187–92.

35. Matthew Hunter, *Wicked Intelligence*, esp. 16–23.

36. Nehemiah Grew, *Musæum Regalis Societatis; or, A Catalogue & Description of the Natural and Artificial Rarities Belonging to the Royal Society* (London, 1681), 204 and tab. 16. For the identity of the fruit, referring to Grew's account, see Commission des Sciences et des Arts, *Description de l'Égypte* [. . .], 2nd ed., 24 vols. (Paris, 1821–29), 19:16.

37. Grew, *Musæum Regalis Societatis*, sig. [A4r]; Richard Foster Jones, "Science and English Prose Style in the Third Quarter of the Seventeenth Century," *PMLA* 45, no. 4 (1930): 977–1009.

38. Peter Galison, "Descartes's Comparisons: From the Invisible to the Visible," *Isis* 75, no. 2 (1984): 311–26, esp. 324; Stephen Gaukroger, "Descartes's Early Doctrine of Clear and Distinct Ideas," *Journal of the History of Ideas* 53, no. 4 (1992): 585–602; M. Jones, *Good Life in the Scientific Revolution*, 55–88, esp. 70.

39. Courtney Weiss Smith, "Rhyme and Reason in John Wilkins's Philosophical Language Scheme," *Modern Philology* 115, no. 2 (2017): 183–212. See also Walmsley, *Locke's Essay*, esp. 108–17; William T. Lynch, *Solomon's Child: Method in the Early Royal Society of London* (Stanford, CA: Stanford University Press, 2001), 152–53.

40. The classic statement of this position, much emulated in works dealing both with scientific rhetoric and images, is Stephen Shapin, "Pump and Circumstance: Robert Boyle's Literary Technology," *Social Studies of Science* 14, no. 4 (1984): 481–520.

41. I owe this point to Stephen Gaukroger, who brought it to my attention during the question-and-answer session at a presentation I gave at the British Society for the History of Science Annual Conference in Aberdeen, July 2010.

42. Writing on the role of rhetoric in these fields is still for the most part carried out in scholarship that deals with literature. For surveys dealing with the interconnections between rhetoric and other disciplines dealing with the emotions, see Gail Kern Paster, Katherine Rowe, and Mary Floyd-Wilson, eds., *Reading the Early Modern Passions: Essays in the Cultural History of Emotion* (Philadelphia: University of Pennsylvania Press, 2005); Stephen Pender and Nancy S. Struever, eds., *Rhetoric and Medicine in Early Modern Europe* (London: Routledge, 2012); Mark Robson, *The Sense of Early Modern Writing: Rhetoric, Poetics, Aesthetics* (Manchester, UK: Manchester University Press, 2006). On the place of rhetoric in early modern political thought, see Quentin Skinner, *Reason and Rhetoric in the Philosophy of Hobbes* (Cambridge: Cambridge University Press, 1996). On Bernard Lamy, see Thomas Hobbes and Bernard Lamy, *The Rhetorics of Thomas Hobbes and Bernard Lamy*, ed. and with an introduction by John T. Harwood (Carbondale: Southern Illinois University Press, 1986).

43. The study of the visual culture of science now constitutes a distinct subfield in the history of science. As Alex Marr has pointed out in his review article "Knowing Images," the field originated principally through the appropriation of art-historical methods by historians of science,

prompted by the publication of Alpers's *Art of Describing* in 1983. Art-historical methods are still important to the study of scientific images, as instanced by Daniela Bleichmar's survey of the visual culture of science in the early modern Spanish Empire, *Visible Empire: Botanical Expeditions and Visual Culture in the Hispanic Enlightenment* (Chicago: University of Chicago Press, 2012). That said, several scholars have grown uncomfortable with using methods intended for the interpretation of artistic images, noting the enormous number and variety of scientific images that defy such categorization. See for instance Horst Bredekamp, Vera Dünkel, and Birgit Schneider, eds., *The Technical Image: A History of Styles in Scientific Imagery* (Chicago: University of Chicago Press, 2015). Despite such differences, studies of the history of scientific images nevertheless—almost by definition—treat the visual as a distinct category of evidence, to be analyzed apart from scientific texts. In some cases, this tendency almost amounts to treating the visual as an unchanging, transhistorical category, much the same now as it was in the early modern period. See Florike Egmond, *Eye for Detail: Images of Plants in Art and Science, 1550–1630* (London: Reaktion Books, 2017). An exception to this tendency is Sachiko Kusukawa's *Picturing the Book of Nature: Image, Text, and Argument in Sixteenth-Century Human Anatomy and Medical Botany* (Chicago: University of Chicago Press, 2012), which seeks to explain scientific images alongside the texts in which they are often to be found and through the philosophical arguments they were intended to serve.

44. Some recent examples include Claire Preston, *The Poetics of Scientific Investigation in Seventeenth-Century England* (Oxford: Oxford University Press, 2016); Courtney Weiss Smith, *Empiricist Devotions: Science, Religion, and Poetry in Early Eighteenth-Century England* (Charlottesville: University of Virginia Press, 2016); and Frédérique Aït-Touati, *Fictions of the Cosmos: Science and Literature in the Seventeenth Century* (Chicago: University of Chicago Press, 2011).

45. Ofer Gal and Raz Chen-Morris, *Baroque Science* (Chicago: University of Chicago Press, 2012), 9–11.

46. On Newtonian teleology, see Robert H. Hurlbutt III, *Hume, Newton, and the Design Argument* (Lincoln: University of Nebraska Press, 1965), 16; and Neal C. Gillespie, "Natural History, Natural Theology, and Social Order: John Ray and the 'Newtonian Ideology,'" *Journal of the History of Biology* 20, no. 1 (1987), 37.

Chapter One

1. Fernando Vidal and Bernhard Kleeberg, "Introduction: Knowledge, Belief, and the Impulse to Natural Theology," *Science in Context* 20, no. 3 (2007): 387–96.

2. John Wilkins, *Of the Principles and Duties of Natural Religion* (London, 1678), 39.

3. Scott Mandelbrote, "The Uses of Natural Theology in Seventeenth-Century England," *Science in Context* 20, no. 3 (2007): 458–59. These thinkers are often referred to as the "Cambridge Platonists." For an account of their thought, see Frederick C. Beiser, *The Sovereignty of Reason: The Defense of Rationality in the Early English Enlightenment* (Princeton, NJ: Princeton University Press, 1996), 134–83. On their responses to Descartes, see 172–74.

4. John Ray, *The Wisdom of God Manifested in the Works of Creation* (London, 1691), preface, sig. [A7r].

5. As Dmitri Levitin has recently pointed out, some thinkers associated with the early Royal Society pursued varieties of physico-theology that did not make design or order in nature their main evidence for the existence and attributes of God. It was only by the end of the seventeenth century that the genre began to stabilize, becoming almost synonymous with the design argu-

ment. See Dmitri Levitin, "Rethinking English Physico-Theology: Samuel Parker's *Tentamina de Deo* (1665)," *Early Science and Medicine* 19, no. 1 (2014): 28–75. In chapter 2, we shall discuss at length another work of physico-theology that does not deal directly with the evidence of design in nature—Robert Boyle's attempt to demonstrate the physical possibility of the resurrection of the body.

6. Walter Charleton, *The Darknes of Atheism Dispelled by the Light of Nature, a Physico-Theologicall Treatise* (London, 1652), 66–67.

7. Neal C. Gillespie, "Natural History, Natural Theology, and Social Order: John Ray and the 'Newtonian Ideology,'" *Journal of the History of Biology* 20, no. 1 (1987): 29–33; Mandelbrote, "Uses of Natural Theology," 454–56.

8. Gillespie, "Natural History, Natural Theology, and Social Order," 33, 41; Mandelbrote, "Uses of Natural Theology," 468–69 and 473, where he describes the design argument as one that "met the need for self-evidence in matters of religion."

9. Richard S. Westfall, *Science and Religion in Seventeenth-Century England* (1958; repr., Ann Arbor: University of Michigan Press, 1973), 68–69. For more recent suggestions that physico-theology was a matter of belief rather than knowledge, see Rhodri Lewis, *William Petty on the Order of Nature: An Unpublished Manuscript Treatise* (Tempe: Arizona Center for Medieval and Renaissance Studies, 2012), 31; and Brian W. Ogilvie, "Insects in John Ray's Natural History and Natural Theology," in *Zoology in Early Modern Culture: Intersections of Science, Theology, Philology, and Political and Religious Education*, ed. Karl A. E. Enenkel and Paul J. Smith (Boston: Brill, 2014), 257.

10. Jonathan Sheehan, "Thomas Hobbes, D.D.: Theology, Orthodoxy, and History," *Journal of Modern History* 88, no. 2 (2016): 260–61; Peter Harrison, *The Territories of Science and Religion* (Chicago: University of Chicago Press, 2015), esp. 83–116.

11. Quentin Meillassoux, *After Finitude: An Essay on the Necessity of Contingency*, trans. Ray Brassier (London: Bloomsbury, 2012), 1–9.

12. On the early publication history of Ray's *Wisdom of God* in England, see William Derham, *Select Remains of the Learned John Ray* (London, 1760), 63–64. On the translation and publication of Ray's work in continental Europe, see Ogilvie, "Insects in John Ray's Natural History," 241.

13. Henry Guerlac and Margaret C. Jacob, "Bentley, Newton, and Providence: The Boyle Lectures Once More," *Journal of the History of Ideas* 30, no. 3 (1969): 309.

14. The first Boyle lecturers were given by Richard Bentley over the course of 1692. They were published in the same year and in several editions thereafter. In 1699, the whole set of eight sermons was published as Richard Bentley, *The folly and unreasonableness of atheism* [. . .] (London, 1699). Samuel Clarke gave his lectures in 1704, publishing them as Samuel Clarke, *A Demonstration of the Being and Attributes of God* [. . .] *Being the Substance of Eight Sermons Preach'd at the Cathedral-Church of St Paul* (London, 1705). William Derham gave his lectures in 1712, publishing them in the following year. See William Derham, *Physico-Theology; or, A Demonstration of the Being and Attributes of God, from his Works of Creation* (London, 1713). On the relationship between Derham's *Physico-Theology* and Ray's work, see Charles Raven, *John Ray, Naturalist: His Life and Works* (Cambridge: Cambridge University Press, 1942), 452.

15. Ray, *Wisdom of God*, preface, sig. [A7r]; Derham, *Physico-Theology*, 2–3.

16. Margaret J. Osler, "Whose Ends? Teleology in Early Modern Natural Philosophy," *Osiris* 16 (2001): 153–55. For Aristotle's teleological account of motion, see Aristotle, *Physics, Volume I: Books 1–4*, trans. P. H. Wicksteed and F. M. Cornford (Cambridge, MA: Harvard University

Press, 1957), 3.1–3.190–215. For Aristotle's teleological account of lightness and heaviness, including his discussion of what we would call gravitation, see Aristotle, *On the Heavens*, trans. W. K. C. Guthrie (Cambridge, MA: Harvard University Press, 1939), 4.3.310–50.

17. Robert Boyle, *The Origine of Formes and Qualities* (1666), in Robert Boyle, *The Works of Robert Boyle*, ed. Michael Hunter and Edward B. Davis, 14 vols. (London: Pickering and Chatto, 1999–2000), 5:351–52. See also Robert Boyle, *A Free Enquiry into the Vulgarly Receiv'd Notion of Nature* (1686), in Boyle, *Works*, 10:458–59. On Descartes, see Helen Hattab, *Descartes on Forms and Mechanisms* (Cambridge: Cambridge University Press, 2009), 17–20.

18. Francis Bacon, *The New Organon*, ed. Lisa Jardine and Michael Silverthorne (Cambridge: Cambridge University Press, 2000), bk. 1, aphorism 48, p. 44. See also Peter Harrison, "Physico-Theology and the Mixed Sciences: The Role of Theology in Early Modern Natural Philosophy," in *The Science of Nature in the Seventeenth Century: Patterns of Change in Early Modern Natural Philosophy*, ed. Peter R. Anstey and John A. Schuster (Dordrecht, Neth.: Springer, 2005), 165–83.

19. Ray, *Wisdom of God*, 127–29; Robert Boyle, *A Disquisition about the Final Causes of Natural Things* (1688), in Boyle, *Works*, 11:86–87, 96–97, 120; John Hedley Brooke, "'Wise Men Nowadays Think Otherwise': John Ray, Natural Theology and the Meanings of Anthropocentrism," *Notes and Records of the Royal Society* 54, no. 2 (2000): 199–213.

20. It is crucial to note that neither Epicurus nor his intellectual descendants in the seventeenth century were atheists in the sense that we understand the term today. The crucial issue at stake in the debate over Epicurean atomism was not the existence of God but rather the possibility that the world could be shown to function on its own, without divine design or intervention. On the reception of Epicurean atomism in the second half of the seventeenth century, see Victor Nuovo, *John Locke: The Philosopher as Christian Virtuoso* (Oxford: Oxford University Press, 2017), 59–86; Margaret J. Osler, *Divine Will and the Mechanical Philosophy: Gassendi and Descartes on Contingency and Necessity in the Created World* (Cambridge: Cambridge University Press, 1994), 36–77; Catherine Wilson, *Epicureanism at the Origins of Modernity* (Oxford: Oxford University Press, 2017). On atheism in the seventeenth century—and especially the point that believing in God was then not enough to avoid the charge of atheism—see Sheehan, "Theology, Orthodoxy, and History," esp. 262–63.

21. Boyle, *Final Causes*, 81–82, 86; Ray, *Wisdom of God*, 13–20. Richard Bentley, *A Confutation of Atheism from the Origin and Frame of the World: Part 1* (London, 1692), 18.

22. Ray, *Wisdom of God*, 25. Here, Ray took his text from Ralph Cudworth, *The True Intellectual System of the Universe* (London, 1678), 683.

23. For Boyle's position, see Boyle, *Final Causes*, 92–95. Boyle's engagement with Descartes's thought will be discussed in more detail subsequently.

24. Cicero, *On the Nature of the Gods: Academics*, trans. H. Rackham (Cambridge, MA: Harvard University Press, 1933), bk. 2, pp. 216–17. See 232–33 for stars, 262–63 for ears.

25. David Foster, "'In Every Drop of Dew': Imagination and the Rhetoric of Assent in English Natural Religion," *Rhetorica* 12, no. 3 (1994): 312–15. Another crucial source for the physico-theologians was an enumeration of the muscles of the human body offered by the Greco-Roman physician Galen in his *De Fœtuum Formatione* (*On the Formation of the Foetus*). Indeed, Isabel Rivers has shown that this example appeared so often in physico-theological texts that the David Hume targeted it for satirical treatment when dismantling the design argument in his *Dialogues Concerning Natural Religion* (1779). See Isabel Rivers, "'Galen's Muscles': Wilkins, Hume, and the Educational Use of the Argument from Design," *Historical Journal* 36, no. 3 (1993): 577–97.

26. Levitin, "Rethinking English Physico-Theology," 65, 73.

27. Immanuel Kant, *Critique of the Power of Judgment*, ed. Paul Guyer, trans. Paul Guyer and Eric Matthews (Cambridge: Cambridge University Press, 2000), 303; Harrison, "Physico-Theology and the Mixed Sciences," 181. An exception to the tendency to conflate different versions of the design argument together is Robert H. Hurlbutt III, *Hume, Newton, and the Design Argument* (Lincoln: University of Nebraska Press, 1965), esp. 81–108.

28. Raven, *John Ray*, 457–69; Gillespie, "Natural History, Natural Theology, and Social Order," 41.

29. Henry More, *An Antidote against Atheisme; or, An Appeal to the Natural Faculties of the Minde of Man, whether there be not a God* (London, 1653), 43–104. On the identity of Julius Scaliger, see Raven, *John Ray*, 460. C. A. Patrides mistakes Julius Scaliger for Joseph Justus Scaliger. See C. A. Patrides, *The Cambridge Platonists*, 2nd ed. (Cambridge: Cambridge University Press, 1980), 284.

30. More, *Antidote*, 94–95.

31. Cicero, *Nature of the Gods*, bk. 2, pp. 260–61.

32. More, *Antidote*, 95–96.

33. Ray, *Wisdom of God*, 180. David Foster points out Ray's use of Cicero but does not take note of Ray's reliance on More's version. See Foster, "Imagination and the Rhetoric of Assent," 316–17.

34. More, *Antidote*, 95.

35. Ray, *Wisdom of God*, 175.

36. Cf. More, *Antidote*, 95–96, and Ray, *Wisdom of God*, 95–98.

37. The consensus today is that the lens itself changes shape, rather than the eyeball.

38. William Derham also offered much-expanded versions of the examples given by his predecessors, including both More and Ray. See, for instance, Derham, *Physico-Theology*, 93–94. Where More had been content simply to wonder out loud whether moles had real eyes or not, Derham presented a detailed description based on his own dissections and observations with the microscope. Cf. More, *Antidote*, 88.

39. Claude Perrault, *Memoir's for a natural history of animals containing the anatomical descriptions of several creatures dissected by the Royal Academy of Sciences at Paris*, trans. Alexander Pitfield (London, 1688), 248. Ray faithfully transcribed Pitfield's English, which in turn was a faithful rendition of the passage given in the original French. Cf. [Claude Perrault], *Mémoires pour servir à l'histoire naturelle des animaux*, 2 vols. (Paris, 1671–76), 2:190: "Les particularitez de la structure admirable de cette Paupière, sont de ces choses qui font voir distinctement la sagesse de la Nature entre mille autres dont nous ne voyons point l'artifice, parce que nous ne les connoissons que par des effets, dont nous ignorons les causes: mais il s'agit icy d'une machine, dont toutes les piéces sont visibles, & qu'il ne faut que regarder, pour découvrir les raisons de son mouvement & de son action."

40. John Ray, *The Wisdom of God Manifested in the Works of Creation*, 2nd ed. (London, 1692), 118–19.

41. The comparison between the eyelid and a palisade is one of few such comparisons in either work.

42. René Descartes, *Meditations on First Philosophy with Selections from the Objections and Replies*, trans. Michael Moriarty (Oxford: Oxford University Press, 2008). Descartes's dismissal of final causes comes in the fourth of the *Meditations*, 40. For Gassendi's objection, see 175. In his replies to Pierre Gassendi's objections to the *Meditations*, Descartes gave a more extended account of his thoughts; see 193. For further discussions of Descartes's rejection of teleological

reasoning, see also Matthew L. Jones, *The Good Life in the Scientific Revolution: Descartes, Pascal, Leibniz, and the Cultivation of Virtue* (Chicago: University of Chicago Press, 2006), 227; and Margaret J. Osler, "From Immanent Natures to Nature as Artifice: The Reinterpretation of Final Causes in Seventeenth-Century Natural Philosophy," *Monist* 79, no. 3 (1996): 392–93.

43. Lewis, *William Petty on the Order of Nature*, 22–26.

44. On Boyle's engagement with Descartes's thought, see Edward B. Davis, "'Parcere Nominibus': Boyle, Hooke, and the Rhetorical Interpretation of Descartes," in *Robert Boyle Reconsidered*, ed. Michael Hunter (Cambridge: Cambridge University Press, 1994), 159–63; and Timothy Shanahan, "Teleological Reasoning in Boyle's *Disquisition about Final Causes*," in Hunter, *Robert Boyle Reconsidered*, 178–83.

45. Gassendi made this argument in his objection to Descartes's *Meditations*. See Descartes, *Meditations*, 175.

46. Boyle, *Final Causes*, 89–90.

47. Boyle, *Final Causes*, 97.

48. Boyle, *Final Causes*, 92–93.

49. Boyle, *Final Causes*, 89.

50. Boyle, *Final Causes*, 102–3, 142 (on the eyes of fishes), and 98, 126 (on diseases). At the end of the work, Boyle also included a supplement detailing "Some Uncommon Observations about Vitiated Sight," 153–67.

51. Robert Boyle, *The Christian Virtuoso* (London, 1690–91), in Boyle, *Works*, 11:297.

52. On the early modern belief that laws of nature required a metaphysical foundation, see John Henry, "Metaphysics and the Origins of Modern Science: Descartes and the Importance of Laws of Nature," *Early Science and Medicine* 9, no. 2 (2004): 73–114, esp. 109.

53. Timothy Shanahan, "God and Nature in the Thought of Robert Boyle," *Journal of the History of Philosophy* 26, no. 4 (1988): 565–67.

54. Boyle, *Final Causes*, 110–11; Boyle, *Formes and Qualities*, 353–55. See also Osler, "Whose Ends?," 159–64; and Brian W. Ogilvie, "Natural History, Ethics, and Physico-Theology," in *Historia: Empiricism and Erudition in Early Modern Europe*, ed. Gianna Pomata and Nancy G. Siraisi (Cambridge, MA: MIT Press, 2005), 97.

55. Jan W. Wojcik, *Robert Boyle and the Limits of Reason* (Cambridge: Cambridge University Press, 1997), 166–74. See also Boyle, *Formes and Qualities*, 351–52, and *Final Causes*, 110.

56. Boyle, *Final Causes*, 111–12. Cf. Boyle, *Formes and Qualities*, 354.

57. Jonathan Sheehan and Dror Wahrman, *Invisible Hands: Self-Organization and the Eighteenth Century* (Chicago: University of Chicago Press, 2015), 33–35. See Ray, *Wisdom of God*, 32–27 and 86–107, on the instinctual behavior of animals, the building of nests, and the apparent care taken for the success of reproduction. On Grew, see Brian Garrett, "Vitalism and Teleology in the Natural Philosophy of Nehemiah Grew (1641–1712)," *British Journal for the History of Science* 36, no. 1 (2003): 63–81. Garrett notes that in the 1670s, Grew's explanations were markedly more mechanistic than in later works such as the *Cosmologia Sacra*, where he asserted far more strongly the need for a plastic principle of the sort described by Cudworth.

58. Boyle, *Christian Virtuoso*, 297. As Boyle notes, the part of this quotation given in italics is a paraphrase of Isaiah 28:29.

59. Ray, *Wisdom of God*, preface, sig. [A8r]. See also 117, 150, 230.

60. William F. Bynum, "The Anatomical Method, Natural Theology, and the Functions of the Brain," *Isis* 64, no. 4 (1973): 453, 458, 460–66; Garrett, "Vitalism and Teleology," 71–73; Ray, *Wisdom of God*, 88–92, esp. 89.

61. Sheehan and Wahrman, *Invisible Hands*, 36–40.

62. Osler, "Whose Ends?," 164–67; Boyle, *Formes and Qualities*, 354.

63. Peter Anstey, "Boyle on Seminal Principles," *Studies in History and Philosophy of Science Part C: Biological and Biomedical Sciences* 33, no. 4 (2002): 625–27.

Chapter Two

1. Robert Boyle, *A Discourse of Things Above Reason. Inquiring Whether a Philosopher should admit there are any such* (1681), in *The Works of Robert Boyle*, ed. Michael Hunter and Edward B. Davis, 14 vols. (London: Pickering and Chatto, 1999–2000), 9:367.

2. William R. Newman, *Atoms and Alchemy: Chymistry and the Experimental Origins of the Scientific Revolution* (Chicago: University of Chicago Press, 2006), 172–73, 179–80, 180n51. Peter Anstey makes the same point, using the term "molecular corpuscle" to make the same distinction. See Peter Anstey, *The Philosophy of Robert Boyle* (London: Routledge, 2000), 44, 63n28.

3. Boyle, *Things Above Reason*, 377.

4. For Descartes's account of sensation, imagination and memory, see René Descartes, *The World and Other Writings*, ed. Stephen Gaukroger (Cambridge: Cambridge University Press, 1998), 146–49.

5. This work was first published in Latin as Thomas Willis, *Cerebri Anatome, cui accessit Nervorum Descriptio et Usus* (London, 1664). The work was published in English as part of Thomas Willis, *Dr. Willis's Practice of Physick, Being the whole Works of that Renowned and Famous Physician*, trans. Samuel Pordage (London, 1684). The part corresponding to the *Cerebri Anatome* is entitled *The Anatomy of the Brain*, and it has its own pagination. Henceforth it will be referred to as Willis, *Anatomy of the Brain*.

6. This work was first published in Latin as Thomas Willis, *De Anima Brutorum* (Oxford, 1672). Like the *Cerebri Anatome*, it was published in English as part of Willis, *Dr. Willis's Practice of Physick*. *De Anima Brutorum* appeared in this edition, with separate pagination, as Thomas Willis, *Two Discourses Concerning the Soul of Brutes*. Henceforth, it will be referred to as *Soul of Brutes*.

7. Willis, *Anatomy of the Brain*, 46, 75, 87; and Willis, *Soul of Brutes*, 1, 3, 34, 38–44.

8. Willis, *Anatomy of the Brain*, 75, 78.

9. Willis, *Anatomy of the Brain*, 79.

10. Willis, *Anatomy of the Brain*, 76.

11. Willis, *Anatomy of the Brain*, 76.

12. Shortly after the *Cerebri Anatome* came out, the Danish naturalist Nicolas Steno (1638–1686) called Willis's hypothesis into question, characterizing it as a work of the imagination for which observational evidence was sorely lacking. See Nicolas Steno, *Discours de Monsieur Stenon, sur L'Anatomie du Cerveau* (Paris, 1669), 11–12. During the eighteenth century, the influential physician George Cheyne (1671/2–1743) was one of many thinkers who rejected Willis's hypothesis that imperceptible spirits passed through the brain, hinting that those spirits were mere figments of Willis's imagination. See George Cheyne, *The English Malady* (London, 1733), 78. Despite such disagreements, Cheyne and his contemporaries generally accepted Willis's basic contention that the mechanisms of the brain and nervous system were crucial to sensory perception and the transmission of commands from the brain to the rest of the body. Following Newton, Cheyne argued that the presence of an extremely subtle, elastic fluid in the nerves made it possible for them to rapidly transmit ideas to and from the brain. Although he hinted that this

fluid might be immaterial, he nevertheless argued that its successful operation also depended on the tone and texture of the bodily organs (see, e.g., Cheyne, *English Malady*, 88–89). On mechanical accounts of sensation, cognition, and voluntary motion in the eighteenth century, see Alison Muri, "Enlightenment Cybernetics: Communications and Control in the Man-Machine," *Eighteenth Century* 49, no. 2 (2008): 155. On the pervasive influence of such models in the culture of the eighteenth century, see G. S. Rousseau, "Nerves, Spirits, and Fibres: Towards Defining the Origins of Sensibility," in *Studies in the Eighteenth Century III: Papers Presented at the Third David Nichol Smith Memorial Seminar 1973*, ed. R. F. Brissenden and J. C. Eade (Canberra: Australian National University Press, 1976).

13. John Ray, *The Wisdom of God Manifested in the Works of Creation* (London, 1691), 38–40; Nehemiah Grew, *Cosmologia Sacra; or, A Discourse of the Universe as it is the Creature and Kingdom of God* (London, 1701), 37–48. See also William F. Bynum, "The Anatomical Method, Natural Theology, and the Functions of the Brain," *Isis* 64, no. 4 (1973): 460–62.

14. Grew, *Cosmologia Sacra*, 44.

15. Grew, *Cosmologia Sacra*, 42–44, quotation at 43.

16. On Boyle's broad agreement with such physiological accounts of the generation of ideas—if not their precise details—see J. J. MacIntosh, "Perception and Imagination in Descartes, Boyle and Hooke," *Canadian Journal of Philosophy* 13, no. 3 (1983): 334.

17. See Jonathan Sheehan, "Thomas Hobbes, D.D.: Theology, Orthodoxy, and History," *Journal of Modern History* 88, no. 2 (2016), 265–68.

18. Thomas Hobbes, *Leviathan*, ed. Richard Tuck, rev. student ed. (Cambridge: Cambridge University Press, 1996), ch. 12, p. 77, my emphasis. See also ch. 3, p. 23, and ch. 31, p. 250.

19. Willis, *Soul of Brutes*, 39.

20. Boyle, *Things Above Reason*, 415.

21. See Robert Boyle, *The Christian Virtuoso* (London, 1690–91), in Boyle, *Works*, 11:285: "As if, because Rational Spirits are Invisible and Immaterial Beings, all Disquisitions about them must be airy and uncertain Speculations, and, like their Objects, devoid of Solidity and Usefulness. But though among these Ingenious Men there are several, whose Expectations from me I am much more disposed to Gratify, than Disappoint; yet, on such an occasion as this, I must take the liberty to own, That I do not think the Corporeal World, nor the Present State of Things, the Only or the Principal Subjects, that an Inquisitive Man's Pen may be worthily employed about." On Boyle's agreement with Descartes on the limits of the imagination and the possibility of possessing immaterial intuitions, see MacIntosh, "Perception and Imagination," 328, 342.

22. Boyle, *Things Above Reason*, 366–67, 404.

23. Boyle, *Things Above Reason*, 367–69.

24. Jan W. Wojcik, *Robert Boyle and the Limits of Reason* (Cambridge: Cambridge University Press, 1997), ch. 2–4, but esp. pp. 106–8.

25. Wojcik, *Robert Boyle and the Limits of Reason*, 7.

26. On Boyle's approach to miracles, see Peter Harrison, "Newtonian Science, Miracles, and the Laws of Nature," *Journal of the History of Ideas* 56, no. 4 (1995): 535. On the importance of theological voluntarism to the emergence of empiricism, see Margaret J. Osler, "Mixing Metaphors: Science and Religion or Natural Philosophy and Theology in Early Modern Europe," *History of Science* 35, no. 1 (1997): 104–6. On the role of voluntarism in Boyle's thought in particular, see Margaret J. Osler, "The Intellectual Sources of Robert Boyle's Philosophy of Nature: Gassendi's Voluntarism and Boyle's Physico-Theological Project," in *Philosophy, Science, and Religion in England, 1640–1700*, ed. Richard W. F. Kroll, Richard Ashcraft, and Perez Zagorin (Cambridge:

Cambridge University Press, 1992), esp. 185–87. Neither the term *voluntarism* nor its antonym *intellectualism* were in fact used during the seventeenth century to describe theological positions. Peter Harrison has thus questioned the tendency to classify natural philosophers such as Descartes and Boyle as either voluntarists or intellectualists, noting a good deal of fluidity and inconsistency in their positions. See Peter Harrison, "Voluntarism and Early Modern Science," *History of Science* 40, no. 1 (2002): 63–89. Harrison's intervention led John Henry to responded by reasserting the value of the terms, seeking to remind Harrison that the theological distinction between voluntarists and intellectualists corresponded to the distinction between empiricists and rationalists in natural philosophy. See John Henry, "Voluntarist Theology at the Origins of Modern Science: A Response to Peter Harrison," *History of Science* 47, no. 1 (2009): 79–113.

27. Wojcik, *Robert Boyle and the Limits of Reason*, 206–9.

28. Wojcik, *Robert Boyle and the Limits of Reason*, 168–204.

29. Boyle, *Things Above Reason*, 398–99. See also 377, as cited at note 3, for the quotation in the introductory section of this chapter.

30. Grew, *Cosmologia Sacra*, 12. Grew defined infinite divisibility not in absolute terms but rather in relation to human capacities. See p. 11, "For as far as the Whole is Extensible, so far the Parts are also Divisible, both Indefinitely; or as Mathematicians speak, Infinitely: that is, beyond any Human Observation or Conception."

31. This is the interpretation given by Lawrence M. Principe, *The Aspiring Adept: Robert Boyle and His Alchemical Quest* (Princeton, NJ: Princeton University Press, 1998), 194. I regard this reading not as incorrect but rather as incomplete. It gives us an accurate account of one part of Boyle's intended meaning, but not the whole.

32. Boyle, *Things Above Reason*, 399.

33. Alan Chalmers has shown that Boyle asserted the value of such incomplete explanations when dealing more directly with the mechanical philosophy. In his essay "About the Excellency and Grounds of the Mechanical Hypothesis" (1674), for instance, Boyle argued that hypotheses that failed to explain phenomena in terms of first principles could nevertheless provide the mind with partial insights that were both pleasing and useful. See Alan Chalmers, "The Lack of Excellency of Boyle's Mechanical Philosophy," *Studies in History and Philosophy of Science* 24, no. 4 (1993): 556–58.

34. Descartes, *Meditations*, 51–52. J. J. MacIntosh offers another interpretation of Boyle's treatment of Descartes's polygon example, stressing points of agreement between the two philosophers (MacIntosh, "Perception and Imagination," 342).

35. Boyle, *Things Above Reason*, 377.

36. Boyle, *Things Above Reason*, 377.

37. Boyle, *Things Above Reason*, 384–85. Other comparisons to the capacities to the eye are to be found at pp. 380, 382–83, 386, 398–99, 401, 412, 415–16, and 420. In addition, Boyle mobilized similar comparisons in referring to the capacities of the imagination. Excluding the earlier comparison involving many-sided polygons, these are to be found at pp. 385, 403, and 420 (part of the same example cited in the list of "eye" comparisons in this footnote).

38. Boyle, *Things Above Reason*, 377.

39. Boyle, *Things Above Reason*, 377–78. Boyle again indicated that finite quantities imaginable to humans could be used to give an inadequate account of God's infinite perfections later in the work (389).

40. Boyle made this argument even though he understood that there could be no ratio between a finite quantity and an infinite quantity. See Boyle, *Things Above Reason*, 390. Boyle

must also have been aware of Descartes's stance on this point from his close engagement with Descartes's critique of physico-theology.

41. John Ray, *Three Physico-Theological Discourses* (London, 1693), 52.

42. Salvatore Ricciardo, "Robert Boyle on God's 'Experiments': Resurrection, Immortality and Mechanical Philosophy," *Intellectual History Review* 25, no. 1 (2015): 101–2.

43. Robert Boyle, *Some Physico-Theological Considerations about the Possibility of the Resurrection* (1675), in Boyle, *Works*, 8:303–4. For Boyle, the other most pressing objection was the problem of identity. His account of the resurrection depended on demonstrating that the human body does not always consist of the same particles of matter throughout its entire life. Making this case enabled him to propose that God could bring about the resurrection using any particles of matter he saw fit. Boyle thus needed to demonstrate that, despite the changes in its material composition, the identity of the being inhabiting that body would remain the same. See Fernando Vidal, "Brains, Bodies, Selves, and Science: Anthropologies of Identity and the Resurrection of the Body," *Critical Inquiry* 28, no. 4 (2002): 952–56.

44. Boyle, *Possibility of the Resurrection*, 299–300, 309.

45. Boyle, *Possibility of the Resurrection*, 306.

46. Robert Boyle, *The Origine of Formes and Qualities* (1666), in Boyle, *Works*, 5:395.

47. Newman, *Atoms and Alchemy*, 158–59, 190–215.

48. For a general discussion of Boyle's use of this kind of inference, see Newman, *Atoms and Alchemy*, 188–89.

49. Wojcik, *Robert Boyle and the Limits of Reason*, 167–79, esp. 179.

50. It is worth mentioning that, as Daniel Garber has shown, Descartes eventually concluded that natural philosophy was incapable of yielding certainty about how the world had been put together. In the *Principia Philosophiae* (1644), Descartes hinted that God could have created the universe in any number of different, but equally plausible, ways. Thus the natural philosopher could only hold up explanations as plausible explanations for how natural phenomena might function—not as accounts of how they actually functioned. See Daniel Garber, *Reading Cartesian Philosophy through Cartesian Science* (Cambridge: Cambridge University Press, 2001), 121–28, esp. 125–26, 128.

51. Boyle, *Formes and Qualities*. Cf. his remarks in the preface, p. 302, with 355, 396–97. Boyle blurred the lines between plausible hypotheses and matters of fact on other occasions. Consider, for instance, his attempts to persuade contemporaries that the "spring of the air" was an observable fact, rather than a mere hypothesis. See Steven Shapin and Simon Schaffer, *Leviathan and the Air-Pump: Hobbes, Boyle, and the Experimental Life* (Princeton, NJ: Princeton University Press, 1985), 50, 213.

52. For Boyle's discussion of palingenesis experiments, see Boyle, *Possibility of the Resurrection*, 303. Michael Hunter has shown that there are significant overlaps between *Possibility of the Resurrection* and Boyle's earlier "Essay of the Holy Scriptures." Ricciardo has shown, however, that discussions of palingenesis play a much larger role in the earlier text than they do in the version of *Possibility of the Resurrection* published by Boyle in 1675. See Michael Hunter, *Robert Boyle, 1627–91: Scrupulosity and Science* (Woodbridge, UK: Boydell Press, 2000), 32; and Ricciardo, "Resurrection, Immortality and Mechanical Philosophy," 106.

53. Boyle, *Possibility of the Resurrection*, 307–8.

54. Boyle, *Possibility of the Resurrection*, 311.

55. Boyle, *Possibility of the Resurrection*, 310–11.

56. Scott Mandelbrote, "The Uses of Natural Theology in Seventeenth-Century England," *Science in Context* 20, no. 3 (2007): 468.

57. For Grew's position, see Harrison, "Newtonian Science, Miracles, and the Laws of Nature," 540–41. See also Grew, *Cosmologia Sacra*, 194–204. Ray's position on miracles and portents fell somewhere between those of Grew and Boyle. In his discussion of the earthquake that took place in Jamaica in 1692, he indicated that such occurrences involved natural causes. Yet he also suggested that God's special intervention was required to make them happen at the right moment. See Ray, *Three Physico-Theological Discourses*, 208.

58. Grew, *Cosmologia Sacra*, 17.

59. John Ray to Tancred Robinson, 12 May 1685, in John Ray, *Philosophical Letters between the Late Learned Mr. Ray and several of his Ingenious Correspondents*, ed. William Derham (London, 1718), 183–85, esp. 185.

60. Boyle, *Christian Virtuoso*, 287.

61. Boyle, *Christian Virtuoso*, 287. Lotte Mulligan has also noted that Boyle's remarks here seem to bespeak a metaphysical argument as well as an approach to making things accessible to the imagination. See Lotte Mulligan, "Robert Boyle, 'Right Reason,' and the Meaning of Metaphor," *Journal of the History of Ideas* 55, no. 2 (1994): 254–55.

Chapter Three

1. Edward Lhwyd to John Ray, 30 February 1691/2, in John Ray, *Philosophical Letters between the Late Learned Mr. Ray and several of his Ingenious Correspondents*, ed. William Derham (London, 1718), 256. Until around 1688, Lhwyd gave his name as Lloyd, thereafter using either Lhwyd or Lhuyd. See Brynley F. Roberts, "Lhuyd [Lhwyd; formerly Lloyd], Edward (1659/60?–1709), naturalist and philologist," *Oxford Dictionary of National Biography* (Oxford: Oxford University Press, 2004).

2. Lhwyd to Ray, 30 February 1691/2, in Ray, *Philosophical Letters*, 257.

3. David McGuinness, "Edward Lhuyd's Contribution to the Study of Irish Megalithic Tombs," *Journal of the Royal Society of Antiquaries of Ireland* 12, no. 126 (1996): 62–85, esp. 68.

4. Lhwyd to Ray, 30 February 1691/2, in Ray, *Philosophical Letters*, 256–57.

5. Matthew C. Hunter, *Wicked Intelligence: Visual Art and the Science of Experiment in Restoration London* (Chicago: University of Chicago Press, 2013), esp. 188–221; J. A. Bennett, *The Mathematical Science of Christopher Wren* (Cambridge: Cambridge University Press, 1982), esp. 87–124.

6. William Poole, *The World Makers: Scientists of the Restoration and the Search for the Origins of the Earth* (Oxford: Peter Lang, 2010); Joseph M. Levine, *Dr. Woodward's Shield: History, Science, and Satire in Augustan England* (Ithaca, NY: Cornell University Press, 1977); Alexander Wragge-Morley, "A Strange and Surprising Debate: Mountains, Original Sin and 'Science' in Seventeenth-Century England," *Endeavour* 33, no. 2 (2009): 76–80. Burnet's work was first published as Thomas Burnet, *Telluris Theoria Sacra*, 2 vols. (London, 1681). The somewhat altered English edition comprised four books in two volumes. See Thomas Burnet, *The Theory of the Earth*, 2 vols. (London, 1684–90).

7. Inigo Jones and John Webb, *The most notable Antiquity of Great Britain vulgarly called* STONE-HENG *on Salisbury Plain. Restored by* INIGO JONES *Esquire, Architect Generall to the late* KING (London, 1655).

8. Walter Charleton, *Chorea Gigantum; or, The most Famous Antiquity of* GREAT-BRITAN, *Vulgarly called* STONE-HENG, *Standing on Salisbury Plain, Restored to the* DANES (London, 1663).

9. In one place, for instance, Aubrey compared his own survey of Stonehenge with that produced by Jones, revealing what he took to be considerable inaccuracies on Jones's part. See John Aubrey's "Monumenta Britannica," ff. 62–65, Top. gen. c. 24–5, Bodleian Library Manuscripts. The most recent and authoritative account of Aubrey's interpretation is Kelsey Jackson Williams, *The Antiquary: John Aubrey's Historical Scholarship* (Oxford: Oxford University Press, 2016), 20–45. See also Stuart Piggott, *Antiquity Depicted: Aspects of Archaeological Illustration* (London: Thames and Hudson, 1978).

10. Michael Hunter, "The Royal Society and the Origins of British Archaeology: I," *Antiquity* 45, no. 178 (1971): 117–18.

11. I. Jones and Webb, *Most Notable Antiquity*, 90. See also p. 96, "But, *Architecture* depending upon demonstration, not fancy, the fictions of *Mythologists* are no further to be embraced, then as not impertinently conducing to prove reall truths."

12. I. Jones and Webb, *Most Notable Antiquity*, 65–89, esp. 76–77.

13. Charleton, *Chorea Gigantum*, 18.

14. I. Jones and Webb, *Most Notable Antiquity*, 55, 63. See also John Webb, *A Vindication of Stone Heng Restored*, 2nd ed. (London, 1725), 23.

15. I. Jones and Webb, *Most Notable Antiquity*, 55.

16. *Oxford English Dictionary Online* (July 2018), s.v. "design, n.," accessed 7 November 2018, http://www.oed.com/view/Entry/50840?isAdvanced=false&result=1&rskey=e1HsIw&.

17. During the European Renaissance, the plan came to be conceived as the starting point of an architectural design consisting of three drawings produced alongside one another—plan, section, and elevation. These drawings had all existed independently for a long time, but during the Renaissance, architects and geometers began to construct them—one from another—using orthographic projection. Thus the plan came to be seen as the first inscription of an idea that would be extrapolated by projective means through the section and the elevation. See Robin Evans, *The Projective Cast: Architecture and Its Three Geometries* (Cambridge, MA: MIT Press, 1995), 107–19; and Alberto Pérez-Gómez and Louise Pelletier, *Architectural Representation and the Perspective Hinge* (Cambridge, MA: MIT Press, 1997), 45–49. This was the sense in which John Evelyn discussed the sequence of architectural drawings in the "Account of Architects and Architecture" added to his translation of Roland Fréart's *Parallèle de l'architecture antique avec la moderne* (Paris, 1650). See Roland Fréart and John Evelyn, *A Parallel of the Antient Architecture with the Modern* [. . .] *To which is added an Account of Architects and Architecture, in an Historical and Etymological Explanation of certain Tearms particularly affected by Architects* (London, 1664), 122.

18. I. Jones and Webb, *Most Notable Antiquity*, 59–64, 70.

19. I. Jones and Webb, *Most Notable Antiquity*, 13, 65. It is worth noting here that, while Jones included drawings of Stonehenge as it actually appeared, these played no role in the argument other than to convince the reader of the accuracy of his architectural reconstructions. For the drawings of Stonehenge as Jones encountered it, see 63 and also Webb, *Vindication of Stone Heng*, 92.

20. I. Jones and Webb, *Most Notable Antiquity*, 67–106, esp. 68–70, 86–87, 101–6.

21. Robert Hooke, *Micrographia; or, Some Physiological Descriptions of Minute Bodies* (London, 1665), Preface, sig. [a1r]; Simon Schaffer, "Newtonian Angels," in *Conversations with Angels:*

Essays towards a History of Spiritual Communication, ed. Joad Raymond (Houndmills, UK: Palgrave Macmillan, 2011), 105–6; Michael Aaron Dennis, "Graphic Understanding: Instruments and Interpretation in Robert Hooke's *Micrographia*," *Science in Context* 3, no. 2 (1989): 337.

22. Joanna Picciotto, "Reforming the Garden: The Experimentalist Eden and *Paradise Lost*," *ELH* 72, no. 1 (2005): 27.

23. Hooke, *Micrographia*, 2, 8.

24. This argument might be seen as an anticipation of Gottfried Leibniz's solution to the question of why God appears to permit the occurrence of evil, presented in his *Essais de Théodicée sur la bonté de Dieu* (1710). In spite of appearances, everything that God had created was perfect, but this perfection was only visible to those who could see things in the way that God had originally intended. On Leibniz's theodicy, see Catherine Wilson, *Leibniz's Metaphysics: A Historical and Comparative Study* (Princeton, NJ: Princeton University Press, 2015), 268–303. Leibniz engaged with Hooke's microscopy in his work on the mechanical philosophy. See Wilson, *Leibniz's Metaphysics*, 47.

25. Dennis, "Graphic Understanding," 311–12, 337–38.

26. Hooke, *Micrographia*, 85; Matthew C. Hunter, "Experiment, Theory, Representation: Robert Hooke's Material Models," in *Beyond Mimesis and Convention: Representation in Art and Science*, ed. Roman Frigg and Matthew C. Hunter (Dordrecht, Neth.: Springer, 2010), 201–3.

27. The role of alchemical/chemical experimentation in the theories of matter promoted by leading members of the early Royal Society has been the subject of much recent inquiry. On the role of saline chemistry in Nehemiah Grew's thought, see Anna Marie Roos, "Nehemiah Grew (1641–1712) on the Saline Chymistry of Plants," *Ambix* 54, no. 1 (2007): 51–68. Robert Boyle also took a close interest in what the processes of crystallization might reveal about the interactions between the invisible aggregate corpuscles he held responsible for the production of visible substances. See Margaret J. Osler, "Whose Ends? Teleology in Early Modern Natural Philosophy," *Osiris* 16 (2001): 166. For a broader survey of the place of alchemy in Boyle's matter theory, see William R. Newman, *Atoms and Alchemy: Chymistry and the Experimental Origins of the Scientific Revolution* (Chicago: University of Chicago Press, 2006).

28. Hooke, *Micrographia*, 81–88.

29. Hooke, *Micrographia*, 88.

30. Hooke, *Micrographia*, 88.

31. Hooke, *Micrographia*, 92.

32. Robert Hooke, "Figures Observ'd in Snow by Mr. Hook," Royal Society Register Book, vol. 2, p. 62. I first encountered this source and the observation about the incisions in the drawing in Matthew C. Hunter, "Robert Hooke Fecit: Making and Knowing in Restoration London" (PhD diss., University of Chicago, 2007), 134–37. He has subsequently discussed it in Matthew Hunter, *Wicked Intelligence*, 49–50.

33. Hooke, *Micrographia*, 91, "In all which I observ'd, that if they were of any regular Figures, they were always branched out with six principal branches."

34. Hooke, *Micrographia*, 91. See also Robert Hooke, "Observables in the six-branch'd Figures in frozen Urine. By Mr. Hook, Decembr. the 10th 1662," Royal Society Register Book, vol. 2, p. 61.

35. Hooke, *Micrographia*, 90. See also Hooke, "Observables in frozen Urine," Royal Society Register Book, vol. 2, p. 61.

36. Hooke, *Micrographia*, Preface, sig. [a2v].

37. Peter N. Miller, "Description Terminable and Interminable: Looking at the Past, Nature,

and Peoples in Peiresc's Archive," in *Historia: Empiricism and Erudition in Early Modern Europe*, ed. Gianna Pomata and Nancy G. Siraisi (Cambridge, MA: MIT Press, 2005), 390.

38. Svetlana Alpers, *The Art of Describing: Dutch Art in the Seventeenth Century* (Chicago: University of Chicago Press, 1983), 73.

39. John Ray to Hans Sloane, 14 December 1698, in John Ray, The *Correspondence of John Ray*, ed. Edwin Lankester (London: Ray Society, 1848), 349. The reviewed work is Paul Hermann, *Paradisus Batavus, Continens Plus Centum Plantas* (Leiden, 1698).

40. John Ray and Francis Willughby, *The Ornithology of Francis Willughby* [. . .] *In Three Books, Wherein All the* BIRDS *Hitherto Known, being reduced into a Method sutable to their Natures, are accurately described* (London, 1678), 137. Ray gave similar advice to Edward Lhwyd in a discussion about ferns. See John Ray to Edward Lhwyd, 20 August 1689, in John Ray, *Further Correspondence of John Ray*, ed. Robert W. T. Gunther (London: Ray Society, 1928), 198.

41. Nehemiah Grew, *Musæum Regalis Societatis; or, A Catalogue & Description of the Natural and Artificial Rarities Belonging to the Royal Society* (London, 1681), preface, sig. [A4v]. For discussions of this defamiliarizing gesture in the context of literary studies, see Michel Beaujour, "Some Paradoxes of Description," *Yale French Studies* 61 (1981): 34–36. See also Walter Bernhart, "Functions of Description in Poetry," in *Description in Literature and Other Media*, ed. Werner Wolf and Walter Bernhart (Amsterdam: Rodopi, 2007).

42. Francis Bacon, *The New Organon*, ed. Lisa Jardine and Michael Silverthorne (Cambridge: Cambridge University Press, 2000), bk. 1, aphorism 48, 44. See also Osler, "Whose Ends?," 155.

43. Grew, *Musæum Regalis Societatis*, preface, sig. [A4v].

44. A notable exception to this inattentiveness is to be found in William F. Bynum, "The Anatomical Method, Natural Theology, and the Functions of the Brain," *Isis* 64, no. 4 (1973): 445–46.

45. The ancient medic Galen made the discovery of the "uses" of the body central to the work of anatomy in his *De Usu Partium Corporis Humani*. For an English translation, see Galen, *On the Usefulness of the Parts of the Body*, ed. and with an introduction by Margaret T. May (Ithaca, NY: Cornell University Press, 1968).

46. Peter Distelzweig, "'Mechanics' and Mechanism in William Harvey's Anatomy: Varieties and Limits," in *Early Modern Medicine and Natural Philosophy*, ed. Peter Distelzweig, Benjamin Goldberg, and Evan R. Ragland (Dordrecht, Neth.: Springer, 2016), 128–29; Don Bates, "*Machina Ex Deo*: William Harvey and the Meaning of Instrument," *Journal of the History of Ideas* 61, no. 4 (2000): 577–93.

47. Domenico Bertoloni Meli, *Mechanism, Experiment, Disease: Marcello Malpighi and Seventeenth-Century Anatomy* (Baltimore: Johns Hopkins University Press, 2011), 355–59.

48. This arrangement is to be found in several anatomical works from the period and sometimes appears in the titles of works, e.g., Thomas Willis, *Cerebri Anatome, cui accessit Nervorum Descriptio et Usus* (London, 1664). I have not been able to locate any scholarly work that deals directly with the emergence of the "description and use" format in publications concerning scientific and navigational instruments. An early example of such a work in English is William Barlow, *The Navigators Supply* [. . .] *With the Description and use of Diverse Instruments* (London, 1598). Around the time Grew and Willis were active, the "description and use" format was still being put to widespread use. See for instance John Brown, *The Description and Use of the Triangular Quadrant* (London, 1671), or William Hunt, *A Mathematical Companion; or, The Description and Use of a New Sliding-Rule* (London, 1697).

49. Nehemiah Grew, *The Anatomy of Plants. With an Idea of a Philosophical History of Plants, and Several Other Lectures, Read before the Royal Society* (London, 1682), 187.

50. Grew, *Anatomy of Plants*, 191.

51. Robert Boyle, *A Disquisition about the Final Causes of Natural Things* (1688), in *The Works of Robert Boyle*, ed. Michael Hunter and Edward B. Davis (London: Pickering and Chatto, 1999–2000), 11:130.

52. Boyle, *Final Causes*, 98.

53. Ernst Cassirer, *The Philosophy of the Enlightenment*, trans. Fritz C. A. Koelln and James P. Pettegrove (Princeton, NJ: Princeton University Press, 1951), 57.

54. Jonathan Loesberg, "Kant, Hume, Darwin, and Design: Why Intelligent Design Wasn't Science before Darwin and Still Isn't," *Philosophical Forum* 38, no. 2 (2007): 101–4. See Immanuel Kant, *Critique of the Power of Judgment*, ed. Paul Guyer, trans. Paul Guyer and Eric Matthews (Cambridge: Cambridge University Press, 2000), 231–84, esp. 266.

55. John Ray to Martin Lister, 19 December 1674, in Ray, *Philosophical Letters*, 129–30. In this case, Ray's instincts led him astray. In normal cases, tropicbirds have either one or two long tail feathers.

56. See, for example, Hans Sloane to John Ray, 31 January 1684/5, in Ray, *Philosophical Letters*, 174; John Ray to Hans Sloane, 17 July 1696, in Ray, *Correspondence*, 297; John Ray to Hans Sloane, 22 March 1698/99, in Ray, *Correspondence*, 362–63; John Ray to Tancred Robinson, 12 May 1685, in Ray, *Further Correspondence*, 146.

57. Nehemiah Grew, *An idea of a phytological history propounded together with a continuation of the anatomy of vegetables, particularly prosecuted upon roots* (London, 1673), 114–18. See also Roos, "Grew on Saline Chymistry," 54–60. The plant fibers referred to by Grew as "succiferous vessels" are today known as xylem tissue.

58. Roos, "Grew on Saline Chymistry," 62; Martin Lister, "A Letter of Mr. Martyn Lister, Written to the Publisher from York, Januar. 10. 1671/2, Containing an Ingenious Account of Veins by Him Observ'd in Plants, Analogous to Human Veins," *Philosophical Transactions of the Royal Society* 6, no. 79 (1671/2): 3055. Cf. Martin Lister to Henry Oldenburg, 10 January 1671/2, in Marin Lister, *The Correspondence of Dr. Martin Lister (1639–1712): Volume One, 1662–1677*, ed. Anna Marie Roos (Leiden, Neth.: Brill, 2015), Letter 0195, p. 415.

59. This exchange, coordinated as it was by Henry Oldenburg, can be found in Oldenburg's published correspondence. The relevant letters are nos. 2283, Lister to Oldenburg, 26 July 1673; 2324 and 2324a, Grew to Oldenburg, 13 September 1673; 2331, Oldenburg to Lister, 18 September 1673; 2369 and 2369a, Lister to Oldenburg, 25 October 1673; 2403, Grew to Oldenburg, 11 December 1673; and 2610, Grew to Lister, 13 December 1673; all in Henry Oldenburg, *The Correspondence of Henry Oldenburg*, ed. and trans. A. Rupert Hall and Marie Boas Hall, 11 vols. (Madison: University of Wisconsin Press, 1965–86), vol. 10. The letters between Lister and Oldenburg may also be found in the edition of Lister's correspondence edited by Roos; see Lister, *Correspondence*.

60. Lister to Oldenburg, 26 July 1673, p. 96. On the final page of the letter (p. 97), Lister made it clear that he found Grew's conjectures ingenious but unconvincing: "The 3d part concerning ye Vegetation of Roots deduced from & founded upon his Observations in ye 2d part is very well becoming ye Author & very ingenious: but yet I had almost said, res adhuc integra est, as to ye tru notion of Vegetation."

61. "Grew's Comments on Lister's Letter 2283," item 2324a, in Oldenburg, *Correspondence*,

208. By "organical," Grew meant to imply that the holes comprised parts of a system that had a use.

62. Grew referred to Hooke's observations. See Grew, *Anatomy of vegetables*, 91. Cf. Hooke, *Micrographia*, 116. Throughout this (1673) version of the work, Grew made use of the "bubble" metaphor to describe the texture of the pith; see, for instance 65–68 and 91–93. When Grew revised the work for the 1682 edition gathered under the title *The Anatomy of Plants*, he removed the word *bubble* and used the word *bladder* instead to describe the appearance of the pith. See, for instance, Nehemiah Grew, "The Anatomy of Roots," in Grew, *Anatomy of Plants*, 75–78.

63. Grew, *Anatomy of vegetables*, 108–10.

64. Lister to Oldenburg, 26 July 1673, pp. 96–97.

65. "Grew's Comments on Lister's Letter 2283," p. 209.

66. Nehemiah Grew, "The Anatomy of Trunks," in Grew, *Anatomy of Plants* (1682), 120–22. William F. Bynum has noted similar tensions in Willis's *Cerebri Anatome* and Ray's works of natural theology. See Bynum, "Anatomical Method," 450, 460–66. Domenico Bertoloni Meli has noted that Malpighi, Hooke, and Grew all made extensive use of analogies between the textures of plants and those to be found in woven fabrics. See Meli, *Mechanism, Experiment, Disease*, 234–70, esp. 270.

67. Grew, *Anatomy of vegetables*, 98–103. Cf. Grew, "Anatomy of Roots," 79–81.

68. Lorraine Daston and Peter Galison, *Objectivity* (Cambridge, MA: Zone Books, 2007), 55–105, esp. 77–79.

69. Christopher Wren, *Wren's "Tracts" on Architecture and Other Writings*, ed. and with an introduction by Lydia M. Soo (Cambridge: Cambridge University Press, 1998), 154–55; Bennett, *Mathematical Science*, 120–23. For a more detailed exposition of Bennett's case, see J. A. Bennett, "Christopher Wren: The Natural Causes of Beauty," *Architectural History* 15 (1972): 5–22.

70. Wren, *"Tracts" on Architecture*, 177–78.

71. Wren, *"Tracts" on Architecture*, 154.

72. Wren, *"Tracts" on Architecture*, 188.

73. John T. Harwood, "Rhetoric and Graphics in *Micrographia*," in *Robert Hooke: New Studies*, ed. Michael Hunter and Simon Schaffer (Woodbridge, UK: Boydell Press, 1989). esp. 135, 142–44.

Chapter Four

1. John Ray, *Historia Plantarum*, 3 vols. (London, 1686–1704).

2. John Ray to Tancred Robinson, 22 October 1684, in John Ray, *Philosophical Letters between the Late Learned Mr. Ray and several of his Ingenious Correspondents*, ed. William Derham (London, 1718), 169.

3. Sachiko Kusukawa, "The *Historia Piscium* (1686)," *Notes and Records of the Royal Society of London* 54, no. 2 (2000): 192. This article contains an excellent discussion of the financial and material difficulties that stood in the way of producing illustrated works of natural history. See also Charles E. Raven, *John Ray, Naturalist: His Life and Works* (Cambridge: Cambridge University Press, 1942), 353.

4. John Ray to Tancred Robinson, 15 December 1690, in Ray, *Philosophical Letters*, 241.

5. John Ray to Richard Waller, 19 May 1691, in John Ray, *Further Correspondence of John Ray*, ed. Robert W. T. Gunther (London: Ray Society, 1928), 99. This letter was subsequently printed as John Ray, "A Letter from That Incomparable Botanist Mr. John Ray, giving an Account of the

Phytographia of Leonard Plukenet, M. D. Lately published. Lond. fol. 1691," *Philosophical Transactions of the Royal Society* 17, no. 194 (1686): 528–30.

6. Brian Vickers, "The Royal Society and English Prose Style: A Reassessment," in *Rhetoric and the Pursuit of Truth: Language Change in the Seventeenth and Eighteenth Centuries*, ed. Brian Vickers and Nancy S. Struever (Los Angeles: Clark Memorial Library, University of California, 1985), 3.

7. Thomas Sprat, *The History of the Royal-Society of London for the Improving of Natural Knowledge* (London, 1667), 112.

8. Sprat, *History of the Royal-Society*, 112.

9. Rhodri Lewis, *Language, Mind and Nature: Artificial Languages in England from Bacon to Locke* (Cambridge: Cambridge University Press, 2007), 155–56; Robert Markley, *Fallen Languages: Crises of Representation in Newtonian England* (Ithaca, NY: Cornell University Press, 1993), 66–67.

10. Sprat, *History of the Royal-Society*, 113.

11. Lewis, *Language, Mind and Nature*, 147–48. The classic statement of the argument that the Royal Society's prose style contributed to the emergence of modern literary aesthetics is Richard Foster Jones, "Science and English Prose Style in the Third Quarter of the Seventeenth Century," *PMLA* 45, no. 4 (1930): 977–1009. For a more recent and wide-ranging discussion of this issue, see John Bender and David E. Wellbery, "Rhetoricality: On the Modernist Return of Rhetoric," in *The Ends of Rhetoric: History, Theory, Practice*, ed. John Bender and David E. Wellbery (Stanford, CA: Stanford University Press, 1990), esp. 10–11.

12. The two earliest and most important reconsiderations of the rhetorical qualities of the Royal Society's language are Stephen Shapin, "Pump and Circumstance: Robert Boyle's Literary Technology," *Social Studies of Science* 14, no. 4 (1984): 481–520; and Vickers, "Royal Society and English Prose Style."

13. Shapin, "Pump and Circumstance," 497.

14. Quentin Skinner, *Reason and Rhetoric in the Philosophy of Hobbes* (Cambridge: Cambridge University Press, 1996), 385, 427–31.

15. Sprat, *History of the Royal-Society*, 416–17. In this passage, Sprat praised Francis Bacon for his prose style, asserting that the great natural philosopher had embellished his prose with the "vast treasure of admirable Imaginations" furnished by his experimental inquiries. Peter Walmsley has shown that elsewhere, in his *Observations into Mons. De Sorbiere's Voyage into England* (London, 1665), Sprat also praised Bacon for his prose, which he found to be "abounding with Metaphors." Walmsley goes on to argue that Sprat found the use of metaphors appropriate to the provisional, hypothetical character of the probable knowledge derived from experimental natural philosophy. See Peter Walmsley, *Locke's Essay and the Rhetoric of Science* (Lewisburg, PA: Bucknell University Press, 2003), 113.

16. William T. Lynch, *Solomon's Child: Method in the Early Royal Society of London* (Stanford, CA: Stanford University Press, 2001), 152–53.

17. Lawrence M. Principe, "Virtuous Romance and Romantic Virtuoso: The Shaping of Robert Boyle's Literary Style," *Journal of the History of Ideas* 56, no. 3 (1995): 377–97, esp. 395.

18. Sachiko Kusukawa, *Picturing the Book of Nature: Image, Text, and Argument in Sixteenth-Century Human Anatomy and Medical Botany* (Chicago: University of Chicago Press, 2012), 103–5, 125–31; Claudia Swan, "The Uses of Realism in Early Modern Illustrated Botany," in *Visualising Medieval Medicine and Natural History*, ed. Jean Givens, Karen M. Reeds and Alain Touwaide (Aldershot, UK: Ashgate, 2006), 248–49. In this piece, Swan also suggests (p. 249) a line of

inquiry that would involve paying more attention to the cognitive effects attributed to images in the early modern period—a suggestion that I take up in this chapter and the one that follows.

19. John Ray and Francis Willughby, *De Historia Piscium* (Oxford, 1686).

20. Martin Lister, *Historiae sive Synopsis Methodicae Conchyliorum*, 5 vols. (London, 1685–92); Leonard Plukenet, *Phytographia, sive Stirpium Illustriorum, et Minus Cognitarium Icones*, 6 vols. (London, 1691–1705).

21. On the production of Lister's images, see Anna Marie-Roos, "The Art of Science: A 'Rediscovery' of the Lister Copperplates," *Notes and Records of the Royal Society* 66, no. 1 (2012): 19–40, esp. 23–34.

22. The practice of using pressed plants as the source for the illustrations in botanical works is discussed in Brian W. Ogilvie, *The Science of Describing: Natural History in Renaissance Europe* (Chicago: University of Chicago Press, 2006), 165–70. In a discussion of another work by Plukenet, the *Almagestum Botanicum* (London, 1696), Ray criticized Plukenet for his dependence on dried plants, remarking that "it is impossible but that a man who relies wholly upon dried specimens of plants (be he never so cunning) should often mistake and multiply." See John Ray to Hans Sloane, 17 July 1696, in John Ray, *The Correspondence of John Ray*, ed. Edwin Lankester (London: Ray Society, 1848), 297.

23. Anna Marie-Roos, "Art of Science," 20–22, 25.

24. John Ray to Richard Waller, 19 May 1691, in Ray, *Further Correspondence*, 99.

25. A minuscule proportion of Lister's plates show comparative dissections, presumably where he found external images of the shells to be unsatisfactory. See, for example the plate in vol. 1 entitled "Harderi Tabulæ Anatomicæ Cochleæ alicujus Terrestris Dorniportæ, earumque explicatio." This is the twenty-second plate of Lister, *Historiae Conchyliorum*, vol. 1 (1685).

26. Barbara Maria Stafford has remarked on Lister's use of these engraved "picture frames," suggesting that in doing so he drew on the visual culture of baroque Jesuit science, exemplified by the work of Athanasius Kircher. See Barbara Maria Stafford, *Artful Science: Enlightenment Entertainment and the Eclipse of Visual Education* (Cambridge, MA: MIT Press, 1994), 231.

27. Nick Grindle, "'No Other Sign or Note than the Very Order': Francis Willughby, John Ray and the Importance of Collecting Pictures," *Journal of the History of Collections* 17, no. 1 (2005): 15–22, at 20.

28. The original source for the whale in this plate is a drawing by Hendrick Goltzius depicting a beached humpback whale. Goltzius drew the whale after it came ashore at Berckhey (in the Netherlands) in 1598. The drawing was engraved by his student Jacob Matham and printed in the same year. The image was widely reproduced in works of natural history, including Carolus Clusius, *Exoticorum Libri Decem* (Leiden, Neth., 1605) and John Johnston, *Historiae Naturalis de Piscibus et Cetis Libri V* (Amsterdam, 1657). The evidence is inconclusive, but Johnston's print could perhaps be the source for the whale in Ray and Willis's *Historia Piscium*. See Susan Dackerman, ed., *Prints and the Pursuit of Knowledge in Early Modern Europe* (Cambridge, MA: Harvard Art Museums, 2011), 216–18. I am indebted to Alex Marr of Cambridge University and one of the anonymous reviewers for pointing out to me the resemblance between Ray's whale and the Matham print.

29. Ray and Willughby, *Historia Piscium*, 27. Ray made this point diplomatically in the *Historia Piscium*, evenhandedly suggesting that cetaceans were closely related both to fish and to quadrupeds: "In universum figura corporis externa piscibus similes sunt, at vero partium internarum & viscerum constitutione & structura cum Quadrupedibus viviparis omnino conveniunt." In his *Synopsis Methodica Stirpium Britannicarum*, 2nd ed. (London, 1696), Ray added

that the penis also served to differentiate cetaceans from fish. This might explain his use of the Matham print, which gives the penis a prominent place, adding a note of specific difference lacking from the images of the dolphin and porpoise, and which would have been invisible if a female whale had been chosen instead. The Royal Society's interest in the problem of classifying cetaceans is perhaps also evidenced by the presence of several whale parts, including a penis, in its repository. Nehemiah Grew, *Musæum Regalis Societatis; or, A Catalogue & Description of the Natural and Artificial Rarities Belonging to the Royal Society* (London, 1681), 81–82.

30. John Ray, "De Methodo Plantarum viri clarissimi D. Augusti Quirini Rivini, [. . .] Epistola ad Joan. Raium, cum ejusdem Responsoria," in Ray, *Synopsis*, 33 (this section has separate pagination and is included at the end of the *Synopsis*). Ray's use of the whale example is discussed in Phillip R. Sloan, "John Locke, John Ray, and the Problem of the Natural System," *Journal of the History of Biology* 5, no. 1 (1972): 1–53, at 38.

31. Sloan, "Locke, Ray, and the Problem," 33–37, 41–42.

32. John Ray, *De Variis Plantarum Methodis Dissertatio Brevis* (London, 1696), 5. Ray's original text is "Essentiae rerum nobis prorus incognitæ sunt. Cùm enim omnis cognitio nostra originem ducat à sensu, nec quicquam de rebus quæ extra nos sunt sciamus, quàm quód facultatem habeant hoc vel illo modo sensus nostros afficiendi, & mediantibus hisce impressionibus, talia vel talia phantasmata in Intellectu excitandi." The translation that I offer here is my own, but I first encountered the passage in Sloan, "Locke, Ray, and the Problem," 43–44.

33. John Ray to Tancred Robinson, 12 May 1685, in Ray, *Philosophical Letters*, 184.

34. Ray, *De Variis Plantarum Methodis Dissertatio Brevis*, 5. I completed this translation with the assistance of Brian W. Ogilvie, to whom I am very grateful. An *incorrect* translation of this passage was offered in 1972 by Sloan in "Locke, Ray, and the Problem," 44. I suspect that this error was simply one of transcription, since Sloan accurately characterizes Ray's meaning despite the mistranslation. However, this mistranslation was carried over, word for word and without proper attribution, by Mary Slaughter in her *Universal Languages and Scientific Taxonomy* (Cambridge: Cambridge University Press, 1982), 212, 261n82.

Ray's original text is "si essentiæ rerum sint formæ immateriales, eas nullo modo sensibus nostris occurrere apud omnes in confesso est: Si verò nihil aliud sint quàm certa quædam proportio & mixtura Principiorum seu minimorum naturalium; cùm minima illa nullo sensu nostro, quantumvis armato aut adjuto seorsim percipi possint, eorum certè figuram aut proportionem sensus nostros fugere & latera necesse est." The problem with Sloan's translation comes in his treatment of the word *quantumvis*, which he renders as "unless." It should be translated as "no matter" in the sense of "no matter how much," and could also be translated in this sense as "however."

35. Sloan, "Locke, Ray, and the Problem," esp. 21–25.

36. John Locke, *An Essay Concerning Human Understanding*, ed. Peter H. Nidditch (Oxford: Oxford University Press, 1975), bk, 2, ch. 8, pp. 132–43, quotation at p. 139.

37. Peter R. Anstey, *John Locke and Natural Philosophy* (Oxford: Oxford University Press, 2011), 156. For a discussion of the primary–secondary qualities distinction in natural philosophy, and the various formulations it has been given, see John J. MacIntosh, "Primary and Secondary Qualities," *Studia Leibnitiana* 8, no. 1 (1976): 88–104, at 94–96.

38. The main difference that I can identify between Locke's position and Ray's is in the treatment of primary and secondary qualities. Ray and Hooke talked about sensory experience as if it could only ever lead to knowledge of what Locke called "secondary qualities"—that is, ideas that did not inhere in the objects of experience but were produced by the operations of external

things on the organs of sensation. This is what I take to be the import of Ray's words in the *Dissertatio*, p. 5, "we know nothing of the things which are outside us except through the ability that they have to make impressions on our senses in some particular way, and by the mediation of these impressions to stir up some phantasm or other in the intellect." Locke, by contrast, insisted that it was possible to know some of the "primary qualities" of things (i.e., qualities actually inhering in those things) by means of sensation. I suspect that the brevity with which Ray discussed his conception of experience can largely account for the difference between these positions. Hooke's similarly brief discussion also glossed over the possibility that some primary qualities might be knowable. See Robert Hooke, "A General Scheme, or Idea of the Present State of Natural Philosophy [. . .]," in Robert Hooke, *The Posthumous Works of Robert Hooke*, ed. Richard Waller (London, 1705), 8–9; Nehemiah Grew, Nehemiah Grew, *Cosmologia Sacra; or, A Discourse of the Universe as it is the Creature and Kingdom of God* (London, 1701), 41 (on the differences between mental ideas, sensations, and external things), 54–55 (on the invisibility of the fundamental particles of matter and the inadequacy of our ideas of them).

39. Locke, *Essay*, bk 3, ch. 11, sec. 21–23, pp. 519–20.

40. Anstey, *Locke and Natural Philosophy*, 51–59; John W. Yolton, *Locke and the Compass of Human Understanding: A Selective Commentary on the "Essay"* (Cambridge: Cambridge University Press, 1970), 42.

41. Locke, *Essay*, bk 3, ch. 11, sec. 21–23, p. 520.

42. Locke, *Essay*, bk. 3, ch. 11, sec. 21–23, pp. 519–20. Cf. Robert Boyle, *The Origine of Formes and Qualities* (1666), in Robert Boyle, *The Works of Robert Boyle*, ed. Michael Hunter and Edward B. Davis, 14 vols. (London: Pickering and Chatto, 1999–2000), 5:322–23.

43. Locke, *Essay*, bk. 3, ch. 11, sec. 23, p. 520.

44. Sloan, "Locke, Ray, and the Problem," 47.

45. Ray and Willughby, *Historia Piscium*, 74. Ray's original text is "Os dentium expers, verum maxillæ cancellatæ tuberculis rhomboidibus asperæ sunt, limæ instar."

46. Carlo Ginzburg, "Ekphrasis and Quotation," *Tijdschrift voor Filosofie* 50, no. 1 (1988): 3–19, at 7.

47. Quintilian, *Institutio Oratoria*, trans. H. E. Butler, 4 vols. (Cambridge, MA: Harvard University Press, 1921), 8.3.61, vol. 3, p. 245.

48. Thomas Hobbes and Bernard Lamy, *The Rhetorics of Thomas Hobbes and Bernard Lamy*, ed. and with an introduction by John T. Harwood (Carbondale: Southern Illinois University Press, 1986), 231. The work was first published as Bernard Lamy, *De l'art de parler* (Paris, 1675). The first English edition was Bernard Lamy, *The Art of Speaking* (London, 1676). Harwood notes that this work was highly influential in England, like its sister work, Antoine Arnauld's *Port-Royal Logic*. The Port-Royal Abbey in Paris was a center of French Jansenism, a theological outlook that had much in common with English latitudinarianism. All references here are from the version presented in Harwood's edition of *The Rhetorics of Thomas Hobbes and Bernard Lamy*, 129–408.

49. There are several excellent works on *enargeia*, all of which discuss this point. See Mark Robson, *The Sense of Early Modern Writing: Rhetoric, Poetics, Aesthetics* (Manchester, UK: Manchester University Press, 2006), 25; Michel Beaujour, "Some Paradoxes of Description," *Yale French Studies* 61 (1981): 28–30, 40–42; Ginzburg, "Ekphrasis and Quotation," 6–7.

50. Quintilian, *Institutio Oratoria*, 8.3.67–70, vol. 3, p. 249.

51. Lamy, *Art of Speaking*, in Hobbes and Lamy, *Rhetorics*, 247.

52. Lamy, *Art of Speaking*, in Hobbes and Lamy, *Rhetorics*, 222, 245.

53. Anthony Grafton, *What Was History? The Art of History in Early Modern Europe* (Cambridge: Cambridge University Press, 2007), 8–12.

54. Ginzburg, "Ekphrasis and Quotation," 7, 10; Plutarch, *Moralia, Volume IV: On the Fame of the Athenians*, trans. Frank Cole Babbitt (Cambridge, MA: Harvard University Press, 1936), 500–501.

55. Horace, *Satires. Epistles. The Art of Poetry*, trans. H. Rushton Fairclough (Cambridge, MA: Harvard University Press, 1926), 480–81; Plutarch, *On the Fame of the Athenians*, 500–501.

56. Rensselaer W. Lee, "*Ut Pictura Poesis*: The Humanistic Theory of Painting," *Art Bulletin* 22, no. 4 (1940): 197–269; on Leonardo, see 251.

57. Gotthold Ephraim Lessing, "Laocoön: An Essay on the Limits of Painting and Poetry (1766)," in *Art in Theory, 1648–1815: An Anthology of Changing Ideas*, ed. Charles Harrison, Paul Wood, and Jason Gaiger (Oxford: Blackwell, 2000), 484–86; Beaujour, "Some Paradoxes of Description," 37–42. See also Wolfgang Ernst, "Not Seeing the *Laocoön*? Lessing in the Archive of the Eighteenth Century," in *Regimes of Description: In the Archive of the Eighteenth Century*, ed. John Bender and Michael Marrinan (Stanford, CA: Stanford University Press, 2005).

58. On the continuing importance of descriptive poetry in eighteenth-century France, and its links with the practices of natural-historical description, see Joanna Stalnaker, *The Unfinished Enlightenment: Description in the Age of the Encyclopedia* (Ithaca, NY: Cornell University Press, 2010), esp. ch. 1 and 2. See also Philippe Hamon and Patricia Baudoin, "Rhetorical Status of the Descriptive," *Yale French Studies* 61 (1981): 7–10.

59. John Dryden, "A Parallel Betwixt Painting and Poetry," in Charles A. Du Fresnoy, *De Arte Graphica: The Art of Painting*, trans. John Dryden (London, 1695), xxxiii.

60. Lee, "*Ut Pictura Poesis*," 252–53. For Addison's praise of visual representation and criticism of description, see Joseph Addison, *The Spectator*, ed. Donald F. Bond, 5 vols. (Oxford: Oxford University Press, 1987), vol. 3, no. 411, pp. 535–37, and no. 416, p. 559. For his endorsement of *enargeia*, see no. 416, p. 560.

61. Locke, *Essay*, bk. 2, ch. 11, sec. 1, p. 155.

62. Locke, *Essay*, bk. 2, ch. 11, sec. 2, p. 156.

63. Taking Locke at face value, scholars therefore generally argue that Locke flatly rejected the possibility that the pleasures of wit could play a constructive role in the exercise of reason. See for instance Skinner, *Reason and Rhetoric*, 374.

64. On Locke's recognition of the power of wit, see Roger D. Lund, "Wit, Judgment, and the Misprisions of Similitude," *Journal of the History of Ideas* 65, no. 1 (2004): 58–61.

65. Locke, *Essay*, bk. 2, ch. 11, sec. 2, p. 156.

66. Locke, *Essay*, bk. 2, ch. 11, sec. 2, p. 157. It is worth mentioning here that the ideal of *enargeia* also played an important role in the early philosophy of René Descartes. Stephen Gaukroger has shown that Descartes's early formulation of the doctrine of "clear and distinct ideas"—ideas that he judged to be sufficiently clear to play a useful role in cognition—was decisively shaped by the standard of evidence elaborated in rhetorical accounts of description. See Stephen Gaukroger, "Descartes's Early Doctrine of Clear and Distinct Ideas," *Journal of the History of Ideas* 53, no. 4 (1992): 585–602, esp. 591–96.

67. Walmsley, *Locke's Essay*, 109–11.

68. Nehemiah Grew, "The Anatomy of Plants, Begun with a General Account of Vegetation, Grounded thereupon," in Grew, *The Anatomy of Plants. With an Idea of a Philosophical History*

of Plants, and Several Other Lectures, Read before the Royal Society (London, 1682), bk. 1, pp. 29–30; Robert Hooke, *Micrographia; or, Some Physiological Descriptions of Minute Bodies* (London, 1665), 140; Locke, *Essay*, bk. 2, ch. 10, sec. 5, 151–52.

Chapter Five

1. John Ray, *John Ray's Cambridge Catalogue (1660)*, ed. and trans. P. H. Oswald and C. D. Preston (London: Ray Society, 2011), 125. Cf. John Ray, *Catalogus Plantarum circa Cantabrigiam Nascentium* (Cambridge, 1660), 3: "figurâ deinde stirpium singularium, colore, totâque adeò externâ specie concinnâ & decorâ mirificè plerumque delectabamur."

2. John Ray, *The Wisdom of God Manifested in the Works of Creation*, 80. On the insistence on the pleasure to be had in God's designs, see Lisa M. Zeitz, "Addison's 'Imagination' Papers and the Design Argument," *English Studies* 73, no. 6 (1992): 497.

3. Ray, *Wisdom of God*, 79.

4. Ray, *Wisdom of God*, 208.

5. Immanuel Kant, *Critique of Judgment*, ed. Paul Guyer, trans. Paul Guyer and Eric Matthews (Cambridge: Cambridge University Press, 2000). For Kant's remarks on disinterestedness, see 90–94. Kant argued that any conception of interest in an object was incompatible with making a free aesthetic judgment. He even argued that satisfaction in the goodness of an object would prevent autonomous aesthetic judgment, see 92–94. Therefore, his position directly contradicts the physico-theological argument that goodness, beauty, and pleasure went together. For Kant's briefest statement of the argument that the formation of the concept of purposiveness in an object would prevent autonomous aesthetic judgment, see 105–6. He also claimed, again in contradiction of the physico-theological position, that the concept of perfection in an object—insofar as that perfection consisted in a teleological judgment about it fulfilling its purpose—would stand in the way of aesthetic judgment. See 111–13.

6. Kant, *Critique of Judgment*, 96–97, 99–101.

7. As we saw earlier, Burnet's work was first published as Thomas Burnet, *Telluris Theoria Sacra*, 2 vols. (London, 1681). The somewhat altered English edition comprised four books in two volumes. See Thomas Burnet, *The Theory of the Earth*, 2 vols. (London, 1684–90).

8. William Poole, *The World Makers: Scientists of the Restoration and the Search for the Origins of the Earth* (Oxford: Peter Lang, 2010), 57–59; Alexander Wragge-Morley, "A Strange and Surprising Debate: Mountains, Original Sin and 'Science' in Seventeenth-Century England," *Endeavour* 33, no. 2 (2009): 76–77; Burnet, *Theory*, 1:144. On the "Newtonian" approach to miracles and other divine interventions in the fabric of nature, see chapter 2 and Peter Harrison, "Newtonian Science, Miracles, and the Laws of Nature," *Journal of the History of Ideas* 56, no. 4 (1995): 540–41.

9. Burnet, *Theory of the Earth*, 1:140.

10. Marjorie H. Nicholson, *Mountain Gloom and Mountain Glory: The Development of the Aesthetics of the Infinite* (1959; repr., Seattle: University of Washington Press, 1997), 194–95.

11. John Ray, *Three Physico-Theological Discourses* (London, 1693), 34–45. On the role of the debate over Burnet's hypothesis in work on the water cycle, see Yi-Fu Tuan, *The Hydrologic Cycle and the Wisdom of God: A Theme in Geoteleology* (Toronto: University of Toronto Press, 1968).

12. Ray, *Three Physico-Theological Discourses*, 35–36.

13. William Derham, *Physico-Theology; or, A Demonstration of the Being and Attributes of*

God, from his Works of Creation (London, 1713), 80. See also Richard Bentley, *A Confutation of Atheism from the Structure and Origin of Humane Bodies* (London, 1692), 32–40.

14. Robert Boyle, *Some Considerations Touching the Style of the H[oly] Scriptures* (London, 1661), in Boyle, *Works*, 2:379–488. On the work's composition, see Robert Markley, "'A Close (though Mystick) Connection': Boyle's Defense of the Bible," in Robert Markley, *Fallen Languages: Crises of Representation in Newtonian England* (Ithaca, NY: Cornell University Press, 1993).

15. John T. Harwood, "Science Writing and Rhetorical Science: Boyle and Rhetorical Theory," in *Robert Boyle Reconsidered*, ed. Michael Hunter (Cambridge: Cambridge University Press, 1994), 44.

16. Carlo Ginzburg, "Style as Inclusion, Style as Exclusion," in *Picturing Science, Producing Art*, ed. Caroline A. Jones and Peter Galison (New York: Routledge, 1998), 28–32. See also Bernard Lamy, *The Art of Speaking*, in Thomas Hobbes and Bernard Lamy, *The Rhetorics of Thomas Hobbes and Bernard Lamy*, ed. and with an introduction by John T. Harwood (Carbondale: Southern Illinois University Press, 1986), 321.

17. Boyle, *Style of the Scriptures*. This argumentative procedure is announced and described at pp. 395–403. The example cited is p. 416, "some Ratiocinations of Scriptures remain undiscern'd or misunderstood, because of our unacquaintednesse with the Figurative, and (oftentimes) Abrupt way of Arguing usual amongst the Eastern People, who in their Arguments us'd to leave much to the Discretion and Collection of those they dealt with."

18. Boyle, *Style of the Scriptures*, 398. Boyle also used this argument (as had Saint Augustine) to justify the stylistic differences that are apparent among books of the Bible. See 402–3.

19. Boyle, *Style of the Scriptures*, 448.

20. Boyle, *Style of the Scriptures*, 442.

21. On the etymology of the word *style* and its role in discourses about taste, see Ginzburg, "Style as Inclusion, Style as Exclusion," 35.

22. Boyle, *Style of the Scriptures*, 483. The passage is Heb. 4:12 (AV).

23. Ray, *Wisdom of God*, 125, my emphasis.

24. On the antirhetorical posture, see Richard Foster Jones, "Science and English Prose Style in the Third Quarter of the Seventeenth Century," *PMLA* 45, no. 4 (1930): 983–84.

25. Jan Golinski, *Making Natural Knowledge: Constructivism in the History of Science*, 2nd ed. (Chicago: University of Chicago Press, 2005), 108; Peter Walmsley, *Locke's Essay and the Rhetoric of Science* (Lewisburg, PA: Bucknell University Press, 2003).

26. Walmsley, *Locke's Essay*, 30.

27. Lamy, *Art of Speaking*, in Hobbes and Lamy, *Rhetorics*, 312–18.

28. Joseph Addison, "An Essay on the Georgics," in Virgil, *The Works of Virgil: Containing his Pastorals, Georgics, and Æneis*, trans. John Dryden (London, 1697), sig. [¶3v–¶4r], between pp. 48 and 49.

29. Lamy, *Art of Speaking*, in Hobbes and Lamy, *Rhetorics*, 221–22.

30. Peter Walmsley discusses how John Locke shifted, in the same fashion, between rhetorical registers, opening with an appropriately high style and quickly shifting to a far lower style if it was appropriate to the matter at hand. See Walmsley, *Locke's Essay*, 114–15.

31. As explained in n. 5 to chapter 2, this work was first published in Latin as Thomas Willis, *Cerebri Anatome, cui accessit Nervorum Descriptio et Usus* (London, 1664). The work was published in English as part of Thomas Willis, *Dr. Willis's Practice of Physick, Being the whole*

Works of that Renowned and Famous Physician, trans. Samuel Pordage (London, 1684). The part corresponding to the *Cerebri Anatome* is entitled *The Anatomy of the Brain*, and it has its own pagination. Henceforth it will be referred to as Willis, *Anatomy of the Brain*; quotation here at 41. Cf. Willis, *Cerebri Anatome*, Epistola Dedicatoria, sig. [A3v]. The Latin expression used for "in the Anatomical Court" is "in delubro Anatomico." The word *delubrum* has sacral connotations and may be translated as "shrine" or "sanctuary." It is possible, therefore, that the "court" alluded to is one of the courts of the Temple of Solomon.

32. Lucian of Samosata, *The Works of Lucian of Samosata*, trans. H. W. Fowler and F. G. Fowler, 4 vols. (Oxford: Clarendon Press, 1905), 1:68–69.

33. Willis, *Anatomy of the Brain*, 44.

34. Willis, *Anatomy of the Brain*, 49.

35. Nehemiah Grew, *The Anatomy of Plants. With an Idea of a Philosophical History of Plants, and Several Other Lectures, Read before the Royal Society* (London, 1682), bk. 1, p. 2.

36. Grew, *Anatomy of Plants*, bk. 1, p. 22.

37. John Ray to Hans Sloane, October 26 1698, in John Ray, *The Correspondence of John Ray*, ed. Edwin Lankester (London: Ray Society, 1848), 357–58. Ray attributed this description to the herbalist John Banister, an English botanist and missionary then active in Virginia, who died in 1692. Like every other naturalist, Ray frequently employed descriptions written by others when he found them fit for his purposes. This description met with his complete approval, as he mentioned in this letter.

38. John Ray and Francis Willughby, *Historia Piscium* (Oxford, 1686), 74.

39. Willis, *Anatomy of the Brain*, 55.

40. Willis, *Anatomy of the Brain*, "a net admirably variegated or flourished" (48), "Chambers or Vaults" (49), "two stems" (50), "two out-stretched wings" (54), "a Flood-gate" (60), "a bubble" (64), "transverse strings or cords" (66), "the serpentine chanels of an Alembick" (72), "*Balneo Mariæ*" (73), "as it were with furrows" (75), "the bill of a *Pelican*" (77), "a Cylinder rolled about into an Orb" (83), "four Mole-hills" (88), "the Kings High-way" (90), "Machine or Clock," "distinct Store-houses" (97), "little Tad-stoles or Puffes" (100), "so many little holes in a Honey-comb" (104), "the Chest [. . .] of a musical Organ" (106).

41. Willis, *Anatomy of the Brain*, 83.

42. Lamy, *Art of Speaking*, in Hobbes and Lamy, *Rhetorics*, 324.

43. Lamy, *Art of Speaking*, in Hobbes and Lamy, *Rhetorics*, 219; see also 222, 233, 245.

44. Thomas Sprat, *The History of the Royal-Society of London for the Improving of Natural Knowledge* (London, 1667), 416.

45. Lamy, *Art of Speaking*, in Hobbes and Lamy, *Rhetorics*, 219–22.

46. Lamy, *Art of Speaking*, in Hobbes and Lamy, *Rhetorics*, 233. Lamy here gives "Tum Cruor, & Vulsae labuntur ab aethere plumae" (Virgil, *Aeneid* 11.724); the English, supplied by the editor, John T. Harwood, is from Dryden's translation of Virgil, *Works of Virgil*, 570.

47. Lamy, *Art of Speaking*, in Hobbes and Lamy, *Rhetorics*, 233.

48. Willis, *Anatomy of the Brain*, 43.

49. Steven Shapin, "'A Scholar and a Gentleman': The Problematic Identity of the Scientific Practitioner in Early Modern England," *History of Science* 29, no. 3 (1991): 279–327.

50. Willis, *Anatomy of the Brain*, 79.

51. Phillip R. Sloan, "John Locke, John Ray, and the Problem of the Natural System," *Journal of the History of Biology* 5, no. 1 (1972): 43–44; *Cosmologia Sacra; or, A Discourse of the Universe as it is the Creature and Kingdom of God* (London, 1701), 4; Robert Hooke, "A General Scheme, or

Idea of the Present State of Natural Philosophy [. . .]," in Robert Hooke, *The Posthumous Works of Robert Hooke*, ed. Richard Waller (London, 1705), 8–9.

52. As explained in n. 6 to chapter 2, this work was first published in Latin as Thomas Willis, *De Anima Brutorum* (Oxford, 1672). Like the *Cerebri Anatome*, it was published in English as part of Willis, *Dr. Willis's Practice of Physick. De Anima Brutorum* appeared in this edition, with separate pagination, as Thomas Willis, *Two Discourses Concerning the Soul of Brutes*. Henceforth, it will be referred to as Willis, *Soul of Brutes*; the reference here is at p. 36.

53. Robert Hooke, "Lectures of Light, Explicating Its Nature, Properties, and Effects," in Hooke, *Posthumous Works*, 145.

54. Hooke, "Lectures of Light," 141.

55. Lamy, *Art of Speaking*, in Hobbes and Lamy, *Rhetorics*, 306.

56. Lamy, *Art of Speaking*, in Hobbes and Lamy, *Rhetorics*, 308. Volume editor John T. Harwood notes that Lamy took this example from the so-called *Port-Royal Logic*, another work produced by the Jansenist circle of the Port-Royal Abbey, and that the "excellent Person" was the theologian and philosopher Antoine Arnauld, the leading light of that circle.

57. Adrian Johns, *The Nature of the Book: Print and Knowledge in the Making* (Chicago: University of Chicago Press, 1998), 397; Willis, *Soul of Brutes*, 210. On the dangers of excessive reading, see also Willis, *Soul of Brutes*, 111 (headache) and 144 (nightmares).

58. Johns, *Nature of the Book*, 380–81. Johns quotes here from Robert Boyle's "Account of Philaretus in his Minority," in Michael Hunter, ed., *Robert Boyle: By Himself and His Friends* (London: William Pickering, 1994), 8–9.

59. Boyle, *Style of the Scriptures*, 442.

60. Sorana Corneanu, *Regimens of the Mind: Boyle, Locke, and the Early Modern Cultura Animi Tradition* (Chicago: University of Chicago Press, 2011), 158–60.

61. Jan Golinski, "The Care of the Self and the Masculine Birth of Science," *History of Science* 40, no. 22 (2002), 134–39; Lisa Jardine, "Hooke the Man: His Diary and His Health," in *London's Leonardo: The Life and Work of Robert Hooke*, ed. Jim Bennett et al. (Oxford: Oxford University Press, 2003), 190–93. See also Charles T. Wolfe and Ofer Gal, eds., *The Body as Object and Instrument of Knowledge: Embodied Empiricism in Early Modern Science* (Dordrecht, Neth.: Springer, 2010); Steven Shapin, "Descartes the Doctor: Rationalism and Its Therapies," *British Journal for the History of Science* 33, no. 2 (2000): 131–54; Rob Iliffe, "Isaac Newton: Lucatello Professor of Mathematics," in *Science Incarnate: Historical Embodiments of Natural Knowledge*, ed. Christopher Lawrence and Steven Shapin (Chicago: University of Chicago Press, 1998), 121–55; Sorana Corneanu and Koen Vermeir, "Idols of the Imagination: Francis Bacon on the Imagination and the Medicine of the Mind," *Perspective on Science* 20, no. 2 (2012): 183–206; Charles T. Wolfe and Michaela van Esveld, "The Material Soul: Strategies for Naturalising the Soul in an Early Modern Epicurean Context," in *Conjunctions of Mind, Soul and Body from Plato to the Enlightenment*, ed. Danijela Kambaskovic (Dordrecht, Neth.: Springer, 2014), 371–421; Lotte Mulligan, "Self-Scrutiny and the Study of Nature: Robert Hooke's Diary as Natural History," *Journal of British Studies* 35, no. 3 (1996): 311–42.

62. Ray, *Wisdom of God*, 123, my emphasis. It is worth noting here that Ray's language hints at a rather different conception of the mechanics of the effects of sensation on the brain and senses than that hinted at by Willis. As discussed earlier, these naturalists did not agree among themselves about precisely how the brain was shaped by its encounters with sensory stimuli. They all agreed, however, that the fabric of the brain was somehow altered in those encounters.

63. Despite agreeing on this broad point, scholars disagree about precisely how natural phi-

losophers sought to discipline experience. Stephen Shapin and Simon Schaffer argue that the Royal Society's key members guaranteed status of the "experimental fact" as a special, epistemologically useful category of experience by enmeshing experiment in the social conventions of gentlemanly conduct, thereby producing a new kind of scientific witness. See Stephen Shapin and Simon Schaffer, *Leviathan and the Air-Pump: Hobbes, Boyle, and the Experimental Life* (Princeton, NJ: Princeton University Press, 1985), esp. ch. 2, pp. 22–79. Peter Dear follows a rather different approach, arguing that the mathematical models of scientific experience pursued in the seventeenth century were more significant for the development of the sciences than Boyle's experimentalism. He therefore emphasizes the role of "established disciplinary experience" in giving or taking away credibility from experiential claims. See Peter Dear, *Discipline and Experience: The Mathematical Way in the Scientific Revolution* (Chicago: University of Chicago Press, 1995), 248. A fascinating exception to the emphasis on the disciplining of experience is Simon Schaffer's essay "Regeneration: The Body of Natural Philosophers in Restoration England," in Lawrence and Shapin, *Science Incarnate*, 83–120. Here, Schaffer shows that fellows of the Royal Society argued that special bodily attributes marked them out as "regenerate" and therefore better able to make use of the evidence of their senses.

Conclusion

1. Giorgio Agamben, *Taste*, trans. Cooper Francis (London: Seagull Books, 2017), 3–5.

2. Steven Shapin, "The Sciences of Subjectivity," *Social Studies of Science* 42, no. 2 (2011): 177.

3. The major exception here is Shapin, who also argues that the study of taste and other procedures for producing intersubjective agreement can shed light on the history of science. In recent years, scholars working on the history of food have done a lot to break down the barriers between the practices of gustatory judgment and the emergence of changing standards of objectivity. See especially Steven Shapin, "A Taste of Science: Making the Subjective Objective in the California Wine World," *Social Studies of Science* 46, no. 3 (2016): 436–60; Emma Spary, *Eating the Enlightenment: Food and the Sciences in Paris, 1670–1760* (Chicago: University of Chicago Press, 2012); Viktoria von Hoffmann, *From Gluttony to Enlightenment* (Urbana: University of Illinois Press, 2016).

4. John Locke, *An Essay Concerning Human Understanding*, ed. Peter H. Nidditch (Oxford: Oxford University Press, 1975), bk. 2, ch. 21, sec. 41–43, pp. 258–60. Note especially Locke's definitions of good and evil: "what has an aptness to produce Pleasure in us, is that we call *Good*, and what is apt to produce Pain in us, we call *Evil*."

5. Locke, *Essay*, bk. 2, ch. 21, sec. 43–58, pp. 259–73, especially Locke's conditional articulation of his theory of moral action at sec. 58, pp. 272–73, "were the satisfaction of a Lust, and the Joys of Heaven offered at once to any one's Present possession, he would not balance, or err in the determination of his choice." For a typically disapproving account of Locke's ethics, see J. B. Schneewind, "Locke's Moral Philosophy," in *The Cambridge Companion to Locke*, ed. Vere Chappell (Cambridge: Cambridge University Press, 1994), 199–225.

6. Locke, *Essay*, bk. 2, ch. 21, sec. 61–69, pp. 274–81.

7. Locke, *Essay*, bk. 2, ch. 21, sec. 69, p. 280.

8. Locke, *Essay*, bk. 2, ch. 21, sec. 69, pp. 280–81.

9. Simon Grote, *The Emergence of Modern Aesthetic Theory: Religion and Morality in Enlightenment Germany and Scotland* (Cambridge: Cambridge University Press, 2017), 2–4.

10. Jerome Stolnitz, "On the Origins of 'Aesthetic Disinterestedness,'" *Journal of Aesthetics and Art Criticism* 20, no. 2 (1961): 131–43, esp. 131–32. For a more up-to-date account of the role of disinterestedness in Shaftesbury's moral and aesthetic thought, see Preben Mortensen, "Shaftesbury and the Morality of Art Appreciation," *Journal of the History of Ideas* 55, no. 4 (1994): 631–50, especially 637–38. For a lucid account of Shaftesbury's conception of aesthetic disinterestedness, see Timothy M. Costelloe, *The British Aesthetic Tradition: From Shaftesbury to Wittgenstein* (Cambridge: Cambridge University Press, 2013), 19. For a twenty-first-century critique of Stolnitz's argument, see Miles Rind, "The Concept of Disinterestedness in Eighteenth-Century British Aesthetics," *Journal of the History of Philosophy* 40, no. 1 (2002): 67–87.

11. See Dabney Townsend, "From Shaftesbury to Kant: The Development of the Concept of Aesthetic Experience," *Journal of the History of Ideas* 48, no. 2 (1986): 287–305; M. H. Abrams, "Kant and the Theology of Art," *Notre Dame English Journal* 13, no. 3 (1981): 75–106; Paul Guyer, *A History of Modern Aesthetics*, 3 vols. (Cambridge: Cambridge University Press, 2015), 1:97–113.

12. Francis Hutcheson, *An Inquiry into the Original of Our Ideas of Beauty and Virtue* (London, 1725). On Shaftesbury, see Costelloe, *British Aesthetic Tradition*, 15–16. On Hutcheson, see Peter Kivy, *The Seventh Sense: Francis Hutcheson and Eighteenth-Century British Aesthetics*, 2nd ed. (Oxford: Oxford University Press, 2003), 24–42. On the differences between Shaftesbury's and Hutcheson's attempts to define beauty as an internal faculty or sense, see Guyer, *History of Modern Aesthetics*, vol. 1, 100–101.

13. Joseph Addison, *The Spectator*, ed. Donald F. Bond, 5 vols. (Oxford: Oxford University Press, 1987), vol. 3, no. 411, pp. 537–39. On Addison's debts to Locke, see Costelloe, *British Aesthetic Tradition*, 38–39.

14. Aris Sarafianos, "Pain, Labor, and the Sublime: Medical Gymnastics and Burke's Aesthetics," *Representations* 91, no. 1 (2005): 62–66, 72. On Diderot and Hogarth, see Abigail Zitin, "Thinking Like an Artist: Hogarth, Diderot, and the Aesthetics of Technique," *Eighteenth-Century Studies* 46, no. 4 (2013): 555–70. On Hogarth and natural philosophy/medicine, see Elizabeth Athens, "The Vital Ornament: Natural Philosophy in William Hogarth's *The Analysis of Beauty* (1753)," *Oxford Art Journal* 40, no. 3 (2017): 397–418; and Michael Baridon, "Hogarth's 'Living Machines of Nature' and the Theorisation of Aesthetics," in *Hogarth: Representing Nature's Machines*, ed. David Bindman, Frédéric Ogée, and Peter Wagner (Manchester, UK: Manchester University Press, 2001), 185–201.

15. Brian Cowan, "The Curious Mr. Spectator: Virtuoso Culture and the Man of Taste in the Works of Addison and Steele," *Media History* 14, no. 3 (2008): 275–92; Craig Ashley Hanson, *The English Virtuoso: Art, Medicine, and Antiquarianism in the Age of Empiricism* (Chicago: University of Chicago Press, 2009).

16. Lawrence E. Klein, *Shaftesbury and the Culture of Politeness: Moral Discourse and Cultural Politics in Early Eighteenth-Century England* (Cambridge: Cambridge University Press, 1994), 161–65. See also Michael Heyd, *"Be Sober and Reasonable": The Critique of Enthusiasm in the Seventeenth and Early Eighteenth Centuries* (Leiden, Neth.: Brill, 1995).

17. Anthony Ashley Cooper, Third Earl of Shaftesbury, "A Letter Concerning Enthusiasm to my Lord *****," in Shaftesbury, *Characteristics of Men, Manners, Opinions, Times*, ed. Lawrence E. Klein (Cambridge: Cambridge University Press, 2000), 27. Shaftesbury's position on enthusiasm was complex, and he made (as we shall see) the difficulty of distinguishing between the true and false forms of enthusiasm central to his reflections on the topic. See Heyd, *Critique of Enthusiasm*, 224–25. On the links between enthusiasm and what we would call aesthetic

experience in Shaftesbury's thought, see Sarah Eron, *Inspiration in the Age of Enlightenment* (Newark: University of Delaware Press, 2014), 42–44. On the links between aesthetic experience and virtue, see Klein, *Shaftesbury and the Culture of Politeness*, 35.

18. Costelloe, *British Aesthetic Tradition*, 15–17; Grote, *Emergence of Modern Aesthetic Theory*, 171–73.

19. Shaftesbury, "Letter Concerning Enthusiasm," 27; Heyd, *Critique of Enthusiasm*, 224–26.

20. Shaftesbury, "Letter Concerning Enthusiasm," 15–16.

21. Shaftesbury, "Letter Concerning Enthusiasm," 15–16. Cf. 23–24.

22. As the name suggests, melancholy had generally been identified as a psychosomatic condition arising from an excess of black bile. The ancient physician Galen argued that the black bile was produced by the liver and filtered or cleansed by the spleen. If some breakdown in the system led there to be too much black bile, harmful effects would ensue. The psychosomatic condition known as melancholy was most often attributed to noxious vapors produced by the black bile rising from the lower organs to disrupt the operations of the brain. See Andrew Wear, "The Spleen in Renaissance Anatomy," *Medical History* 21 (1977): 44–45; and Angus Gowland, *The Worlds of Renaissance Melancholy: Robert Burton in Context* (Cambridge: Cambridge University Press, 2006), 63–64. On the overlaps among hypochondria, hysteria, and melancholy around the turn of the eighteenth century, see Anita Guerrini, *Obesity and Depression in the Enlightenment: The Life and Times of George Cheyne* (Norman: University of Oklahoma Press, 2000), 5–6.

23. Willis, *Soul of Brutes*, 200. On changing medical explanations for enthusiasm, see Heyd, *Critique of Enthusiasm*, 191–96; and Lionel Laborie, *Enlightening Enthusiasm: Prophecy and Religious Experience in Early Eighteenth-Century England* (Manchester, UK: Manchester University Press, 2015), 204–19.

24. Anthony Ashley Cooper, Third Earl of Shaftesbury, "Miscellaneous Reflections on the Preceding Treatises and Other Critical Subjects," in Shaftesbury, *Characteristics*, 355.

25. Klein, *Shaftesbury and the Culture of Politeness*, 65.

26. Addison, *Spectator*, vol. 3, no. 417, pp. 562–63. In the very next line, Addison coyly gestured to the conventional distinction between the soul and the bodily parts of the mind, writing that "it would be in vain to enquire, whether the Power of Imagining Things strongly proceeds from any greater Perfection in the Soul, or from any nicer Texture in the Brain of one Man than of another."

27. Addison, *Spectator*, vol. 3, no. 413, pp. 545–46. See also Lisa M. Zeitz, "Addison's 'Imagination' Papers and the Design Argument," *English Studies* 73, no. 6 (1992): 493–502, at 496.

28. Addison, *Spectator*, vol. 3, no. 421, p. 579.

29. Addison, *Spectator*, vol. 3, no. 421, p. 579.

30. Thomas Sydenham, *Processus Integri in Morbis fere Omnibus Curandis* (London, 1692), 5. For the English translation cited here, see Thomas Sydenham, *The Compleat Method of Curing Almost all Diseases* (London, 1694), 5.

31. Both Locke and Willis accounted for dreams in this manner. Unlike Willis, however, Locke did not attempt to specify the precise mechanisms involved, and he expressed less confidence that dreams arose from the body alone. See Locke, *Essay*, bk. 2, ch. 1, sec. 16–17, p. 113. Cf. Willis, *Soul of Brutes*, 93–95.

32. G. J. Barker-Benfield, *The Culture of Sensibility: Sex and Society in Eighteenth-Century Britain* (Chicago: University of Chicago Press, 1996), esp. 23–36. On the broader role of nervous disorders and their therapies in the culture of the eighteenth century, see John Mullan, "Hypochondria and Hysteria: Sensibility and the Physicians," *Eighteenth Century* 25, no. 2 (1984): 141–

74, and G. S. Rousseau, "Nerves, Spirits, and Fibres: Towards Defining the Origins of Sensibility," in *Studies in the Eighteenth Century III: Papers Presented at the Third David Nichol Smith Memorial Seminar 1973*, ed. R. F. Brissenden and J. C. Eade (Canberra: Australian National University Press, 1976).

33. Roy Porter, "Consumption: Disease of the Consumer Society," in *Consumption and the World of Goods*, ed. John Brewer and Roy Porter (London: Routledge, 1993), 58–81; Barker-Benfield, *Culture of Sensibility*, 25–26.

34. John Purcell, *A Treatise of Vapours; or, Hysterick Fits* (London, 1702), 32; George Cheyne, *The English Malady* (London, 1733), 25–29, 48–51.

35. Porter, "Consumption," 64. Cf. Cheyne, *English Malady*, 149–83. The novelist Samuel Richardson suffered from the nervous disorders so characteristic of his time, and he sought treatment from Cheyne. In one of his many letters to Richardson, Cheyne recommended the use of a "Chamber-horse" to mimic the effects of riding while indoors. See George Cheyne to Samuel Richardson, 20 April 1740, in Samuel Richardson, *Correspondence with George Cheyne and Thomas Edwards*, ed. David E. Shuttleton and John A. Dussinger (Cambridge: Cambridge University Press, 2013), 55. See also Cheyne, *The Natural Method of Cureing the Diseases of the Body, and the Disorders of the Mind Depending on the Body* (London, 1742), 301.

36. Joanna Picciotto, *Labors of Innocence in Early Modern England* (Cambridge, MA: Harvard University Press, 2010), 515–23.

37. Guerrini, *Obesity and Depression in the Enlightenment*, 125–26; Cheyne, *Natural Method*, 314–15. Cf. Cheyne, *English Malady*, 26.

38. Francis Fuller, *Medicina Gymnastica; or, A Treatise Concerning the Power of Exercise, With Respect to the Animal Oeconomy* (London, 1705), 191.

39. Cheyne, *Natural Method*, 316.

40. Purcell, *Treatise of Vapours*, 75–76. I first encountered this reference in Picciotto, *Labors of Innocence*, 515.

41. Purcell, *Treatise of Vapours*, 76; Cheyne, *English Malady*, 4–5; Cheyne, *Natural Method*, 78–88; Fuller, *Medicina Gymnastica*, 202–3.

42. Cheyne, *Natural Method*, 82–83. Here, Cheyne's position was ambiguous. At this point in the text, he asserted that the division between the thinking and the laboring parts of humanity was a permanent one, impressed into nature from the outset. Throughout both the *Natural Method* and the *English Malady*, however, Cheyne continually implied that the opposite was the case, suggesting that diet and exercise could drastically alter an individual's capacity for intellectual activity and sensory experience.

43. The two discourses came out together as Jonathan Richardson, *Two Discourses. I. An Essay On the whole Art of Criticism as it relates to Painting.* [. . .] *II. An Argument in behalf of the Science of a Connoisseur* (London, 1719). However, they have separate pagination. I will refer to them as J. Richardson, *Art of Criticism* and J. Richardson, *Science of a Connoisseur.*

44. Carol Gibson-Wood, *Jonathan Richardson: Art Theorist of the English Enlightenment* (New Haven, CT: Yale University Press, 2000), 179–82.

45. Gibson-Wood, *Jonathan Richardson*, 182–97.

46. J. Richardson, *Art of Criticism*, 98–149, esp. 102–3 on fingers and toes.

47. Carol Gibson-Wood, "Jonathan Richardson's 'Hymn to God,'" *Man and Nature* 8 (1989): 81–90, at 87.

48. Jonathan Richardson, "Hymn to God," British Library Add. MS. 10423, f. 20 recto.

49. J. Richardson, "Hymn to God."

50. J. Richardson, *Art of Criticism*, 1–2.

51. J. Richardson, *Science of a Connoisseur*, 159.

52. J. Richardson, *Science of a Connoisseur*, on tobacco and alcohol, 101–2; on the pathologies of consumption in general, 169–71; on disease, 187–89.

53. J. Richardson, *Science of a Connoisseur*, 173–74.

54. J. Richardson, *Science of a Connoisseur*, 202–3.

55. Lorraine Daston and Peter Galison, *Objectivity* (Cambridge, MA: Zone Books, 2007).

Bibliography

Abrams, M. H. "Kant and the Theology of Art." *Notre Dame English Journal* 13, no. 3 (1981): 75–106.

Addison, Joseph. "An Essay on the Georgics." In *The Works of Virgil: Containing his Pastorals, Georgics, and Æneis*, translated by John Dryden, sig. [¶3v–¶¶2r], between 48 and 49. London, 1697.

———. *The Spectator*. Edited by Donald F. Bond. 5 vols. Oxford: Oxford University Press, 1987.

Agamben, Giorgio. *Taste*. Translated by Cooper Francis. London: Seagull Books, 2017. First published 2015 as *Gusto* by Quodlibet (Macerata, It.).

Aït-Touati, Frédérique. *Fictions of the Cosmos: Science and Literature in the Seventeenth Century*. Chicago: University of Chicago Press, 2011.

Alpers, Svetlana. *The Art of Describing: Dutch Art in the Seventeenth Century*. Chicago: University of Chicago Press, 1983.

Anstey, Peter R. "Boyle on Seminal Principles." *Studies in History and Philosophy of Science Part C: Biological and Biomedical Sciences* 33, no. 4 (2002): 597–630.

———. *John Locke and Natural Philosophy*. Oxford: Oxford University Press, 2011.

———. *The Philosophy of Robert Boyle*. London: Routledge, 2000.

Aristotle. *Art of Rhetoric*. Translated by J. H. Freese. Loeb Classical Library 193. Cambridge, MA: Harvard University Press, 1926.

———. *On the Heavens*. Translated by W. K. C. Guthrie. Loeb Classical Library 338. Cambridge, MA: Harvard University Press, 1939.

———. *Physics, Volume I: Books 1–4*. Translated by P. H. Wicksteed and F. M. Cornford. Loeb Classical Library 228. Cambridge, MA: Harvard University Press, 1957.

Athens, Elizabeth. "The Vital Ornament: Natural Philosophy in William Hogarth's *The Analysis of Beauty* (1753)." *Oxford Art Journal* 40, no. 3 (2017): 397–418.

Aubrey, John. John Aubrey's "Monumenta Britannica," ff. 62–65. Top. gen. c. 24–5. Bodleian Library Manuscripts.

Bacon, Francis. *The New Organon*. Edited by Lisa Jardine and Michael Silverthorne. Cambridge: Cambridge University Press, 2000.

Baridon, Michael. "Hogarth's 'Living Machines of Nature' and the Theorisation of Aesthetics."

In *Hogarth: Representing Nature's Machines*, edited by David Bindman, Frédéric Ogée, and Peter Wagner, 85–101. Manchester, UK: Manchester University Press, 2001.

Barker-Benfield, G. J. *The Culture of Sensibility: Sex and Society in Eighteenth-Century Britain.* Chicago: University of Chicago Press, 1996.

Barlow, William. *The Navigators Supply* [. . .] *With the Description and use of Diverse Instruments.* London, 1598.

Bates, Don. "*Machina Ex Deo*: William Harvey and the Meaning of Instrument." *Journal of the History of Ideas* 61, no. 4 (2000): 577–93.

Baumgarten, Alexander Gottlieb. *Aesthetica*. Frankfurt an der Oder, 1750.

———. *Meditationes Philosophicae de Nonullis ad Poema Pertinentibus*. Magdeburg, Ger., 1735.

Beaujour, Michel. "Some Paradoxes of Description." In "Towards a Theory of Description." Special issue, *Yale French Studies* 61 (1981): 27–59.

Beiser, Frederick C. *The Sovereignty of Reason: The Defense of Rationality in the Early English Enlightenment*. Princeton, NJ: Princeton University Press, 1996.

Bender, John, and David E. Wellbery. "Rhetoricality: On the Modernist Return of Rhetoric." In *The Ends of Rhetoric: History, Theory, Practice*, edited by John Bender and David E. Wellbery, 3–39. Stanford: Stanford University Press, 1990.

Bennett, J. A. "Christopher Wren: The Natural Causes of Beauty." *Architectural History* 15 (1972): 5–22.

———. *The Mathematical Science of Christopher Wren*. Cambridge: Cambridge University Press, 1982.

Bentley, Richard. *A Confutation of Atheism from the Origin and Frame of the World: Part 1.* London, 1692.

———. *A Confutation of Atheism from the Structure and Origin of Humane Bodies*. London, 1692.

———. *The folly and unreasonableness of atheism demonstrated from the advantage and pleasure of a religious life, the faculties of humane souls, the structure of animate bodies, & the origin and frame of the world*. London, 1699.

Bernhart, Walter. "Functions of Description in Poetry." In *Description in Literature and Other Media*, edited by Werner Wolf and Walter Bernhart, 129–52. Studies in Intermediality 2. Amsterdam: Rodopi, 2007.

Bleichmar, Daniela. *Visible Empire: Botanical Expeditions and Visual Culture in the Hispanic Enlightenment*. Chicago: University of Chicago Press, 2012.

Boyle, Robert. *The Christian Virtuoso* (London, 1690–91). In Boyle, *Works*, 11:281–327.

———. *A Discourse of Things Above Reason. Inquiring Whether a Philosopher should admit there are any such* (1681). In Boyle, *Works*, 9:361–424.

———. *A Disquisition about the Final Causes of Natural Things* (1688). In Boyle, *Works*, 11:79–167.

———. *A Free Enquiry into the Vulgarly Receiv'd Notion of Nature* (1686). In Boyle, *Works*, 10:437–571.

———. *The Origine of Formes and Qualities* (1666). In Boyle, *Works*, 5:281–491.

———. *Some Considerations Touching the Style of the H[oly] Scriptures* (1661). In Boyle, *Works*, 2:379–488.

———. *Some Physico-Theological Considerations about the Possibility of the Resurrection* (1675). In Boyle, *Works*, 8:294–313.

———. *The Works of Robert Boyle*. Edited by Michael Hunter and Edward B. Davis. 14 vols. London: Pickering and Chatto, 1999–2000.

Bredekamp, Horst, Vera Dünkel, and Birgit Schneider, eds., *The Technical Image: A History of Styles in Scientific Imagery*. Chicago: University of Chicago Press, 2015.

Brooke, John Hedley. *Science and Religion: Some Historical Perspectives*. Cambridge: Cambridge University Press, 1991.

———. "'Wise Men Nowadays Think Otherwise': John Ray, Natural Theology and the Meanings of Anthropocentrism." *Notes and Records of the Royal Society* 54, no. 2 (2000): 199–213.

Brown, John. *The Description and Use of the Triangular Quadrant*. London, 1671.

Burnet, Thomas. *Telluris Theoria Sacra*. 2 vols. London, 1681.

———. *The Theory of the Earth*. 2 vols. London, 1684–90.

Burtt, E. A. *The Metaphysical Foundations of Modern Physical Science*. London: Routledge and Kegan Paul, 1924.

Bynum, William F. "The Anatomical Method, Natural Theology, and the Functions of the Brain." *Isis* 64, no. 4 (1973): 445–68.

Caron, Louis. "Thomas Willis, the Restoration, and the First Works of Neurology." *Medical History* 59, no. 4 (2015): 525–53.

Cassirer, Ernst. *The Philosophy of the Enlightenment*. Translated by Fritz C. A. Koelln and James P. Pettegrove. Princeton, NJ: Princeton University Press, 1951.

Chalmers, Alan. "The Lack of Excellency of Boyle's Mechanical Philosophy." *Studies in History and Philosophy of Science* 24, no. 4 (1993): 541–64.

Charleton, Walter. *Chorea Gigantum; or, The most Famous Antiquity of GREAT-BRITAN, Vulgarly called STONE-HENG, Standing on Salisbury Plain, Restored to the DANES*. London, 1663.

———. *The Darknes of Atheism Dispelled by the Light of Nature, a Physico-Theologicall Treatise*. London, 1652.

Cheyne, George. *The English Malady*. London, 1733.

———. *The Natural Method of Cureing the Diseases of the Body, and the Disorders of the Mind Depending on the Body*. London, 1742.

Cicero. *On the Nature of the Gods: Academics*. Translated by H. Rackham. Loeb Classical Library 268. Cambridge, MA: Harvard University Press, 1933.

Clark, Stuart. *Vanities of the Eye: Vision in Early Modern Culture*. Oxford: Oxford University Press, 2007.

Clarke, Samuel. *A Demonstration of the Being and Attributes of God [. . .] Being the Substance of Eight Sermons Preach'd at the Cathedral-Church of St Paul*. London, 1705.

Clusius, Carolus. *Exoticorum Libri Decem*. Leiden, Neth., 1605.

Commission des Sciences et des Arts. *Description de l'Égypte; ou Recueil des Observations et des Recherches qui ont été faites en Égypte pendant l'expédition de l'armée française*. 2nd ed. 24 vols. Paris, 1821–29.

Corneanu, Sorana. *Regimens of the Mind: Boyle, Locke, and the Early Modern Cultura Animi Tradition*. Chicago: University of Chicago Press, 2011.

Corneanu, Sorana, and Koen Vermeir. "Idols of the Imagination: Francis Bacon on the Imagination and the Medicine of the Mind." *Perspective on Science* 20, no. 2 (2012): 183–206.

Costelloe, Timothy M. *The British Aesthetic Tradition: From Shaftesbury to Wittgenstein*. Cambridge: Cambridge University Press, 2013.

Cowan, Brian. "The Curious Mr. Spectator: Virtuoso Culture and the Man of Taste in the Works of Addison and Steele." *Media History* 14, no. 3 (2008): 275–92.

Cudworth, Ralph. *The True Intellectual System of the Universe*. London, 1678.

Dackerman, Susan. *Prints and the Pursuit of Knowledge in Early Modern Europe*. Cambridge, MA: Harvard Art Museums, 2011.

Daniel, Drew. "Self-Killing and the Matter of Affect in Bacon and Spinoza." In *Affect Theory and Early Modern Texts: Politics, Ecologies, and Form*, edited by Amanda Bailey and Mario Digagni, 89–108. Palgrave Studies in Affect Theory and Literary Criticism. New York: Palgrave Macmillan, 2017.

Daston, Lorraine. "On Scientific Observation." *Isis* 99, no. 1 (2008): 97–110.

Daston, Lorraine, and Peter Galison. *Objectivity*. Cambridge, MA: Zone Books, 2007.

Davis, Edward B. "'Parcere Nominibus': Boyle, Hooke, and the Rhetorical Interpretation of Descartes." In Michael Hunter, *Robert Boyle Reconsidered*, 157–75.

Dear, Peter. *Discipline and Experience: The Mathematical Way in the Scientific Revolution*. Chicago: University of Chicago Press, 1995.

Dennis, Michel Aaron. "Graphic Understanding: Instruments and Interpretation in Robert Hooke's *Micrographia*." *Science in Context* 3, no. 2 (1989): 309–64.

Derham, William. *Physico-Theology; or, A Demonstration of the Being and Attributes of God, from his Works of Creation*. London, 1713.

———. *Select Remains of the Learned John Ray*. London, 1760.

Descartes, René. *Meditations on First Philosophy with Selections from the Objections and Replies*. Translated by Michael Moriarty. Oxford: Oxford University Press, 2008.

———. *The World and Other Writings*. Edited by Stephen Gaukroger. Cambridge Texts in the History of Philosophy. Cambridge: Cambridge University Press, 1998.

Distelzweig, Peter. "'Mechanics' and Mechanism in William Harvey's Anatomy: Varieties and Limits." In *Early Modern Medicine and Natural Philosophy*, edited by Peter Distelzweig, Benjamin Goldberg, and Evan R. Ragland, 117–40. History, Philosophy and Theory of the Life Sciences 14. Dordrecht, Neth.: Springer, 2016.

Doherty, Meghan C. "Discovering the 'True Form': Hooke's *Micrographia* and the Visual Vocabulary of Engraved Portraits." *Notes and Records of the Royal Society* 66, no. 3 (2012): 211–34.

Dryden, John. "A Parallel Betwixt Painting and Poetry." In *De Arte Graphica: The Art of Painting*, by Charles A. Du Fresnoy, translated and introduced by John Dryden, i–lvi. London, 1695.

Egmond, Florike. *Eye for Detail: Images of Plants in Art and Science, 1550–1630*. London: Reaktion Books, 2017.

Ernst, Wolfgang. "Not Seeing the *Laocoön*? Lessing in the Archive of the Eighteenth Century." In *Regimes of Description: In the Archive of the Eighteenth Century*, edited by John Bender and Michael Marrinan, 118–34. Stanford, CA: Stanford University Press, 2005.

Eron, Sarah. *Inspiration in the Age of Enlightenment*. Newark: University of Delaware Press, 2014.

Evans, Robin. *The Projective Cast: Architecture and Its Three Geometries*. Cambridge, MA: MIT Press, 1995.

Foster, David. "'In Every Drop of Dew': Imagination and the Rhetoric of Assent in English Natural Religion." *Rhetorica* 12, no. 3 (1994): 293–25.

Fréart, Roland. *Parallèle de l'architecture antique avec la moderne*. Paris, 1650.

Fréart, Roland, and John Evelyn. *A Parallel of the Antient Architecture with the Modern* [. . .] *To which is added an Account of Architects and Architecture, in an Historical and Etymological Explanation of certain Tearms particularly affected by Architects*. London, 1664.

Fuller, Francis. *Medicina Gymnastica; or, A Treatise Concerning the Power of Exercise, With Respect to the Animal Oeconomy*. London, 1705.

Gal, Ofer, and Raz Chen-Morris. *Baroque Science*. Chicago: University of Chicago Press, 2012.

Galen. *On the Usefulness of the Parts of the Body*. Edited and with an introduction by Margaret T. May. Ithaca, NY: Cornell University Press, 1968.

Galison, Peter. "Descartes's Comparisons: From the Invisible to the Visible." *Isis* 75, no. 2 (1984): 311–26.

Garber, Daniel. *Reading Cartesian Philosophy through Cartesian Science*. Cambridge: Cambridge University Press, 2001.

Garrett, Brian. "Vitalism and Teleology in the Natural Philosophy of Nehemiah Grew (1641–1712)." *British Journal for the History of Science* 36, no. 1 (2003): 63–81.

Gaukroger, Stephen. "Descartes's Early Doctrine of Clear and Distinct Ideas." *Journal of the History of Ideas* 53, no. 4 (1992): 585–602.

———. *The Emergence of a Scientific Culture: Science and the Shaping of Modernity, 1210–1685*. Oxford: Oxford University Press, 2006.

Gibson-Wood, Carol. *Jonathan Richardson: Art Theorist of the English Enlightenment*. Paul Mellon Centre for Studies in British Art. New Haven, CT: Yale University Press, 2000.

———. "Jonathan Richardson's 'Hymn to God.'" *Man and Nature* 8 (1989): 81–90.

Gillespie, Neal C. "Natural History, Natural Theology, and Social Order: John Ray and the 'Newtonian Ideology.'" *Journal of the History of Biology* 20, no. 1 (1987): 1–49.

Ginzburg, Carlo. "Ekphrasis and Quotation." *Tijdschrift voor Filosofie* 50, no. 1 (1988): 3–19.

———. "Style as Inclusion, Style as Exclusion." In *Picturing Science, Producing Art*, edited by Caroline A. Jones and Peter Galison, 27–54. New York: Routledge, 1998.

Glanvill, Joseph. *Plus Ultra; or, The Progress and Advancement of Knowledge since the days of Aristotle*. London, 1668.

Golinski, Jan. "The Care of the Self and the Masculine Birth of Science." *History of Science* 40, no. 22 (2002): 125–45.

———. *Making Natural Knowledge: Constructivism in the History of Science*. 2nd ed. Chicago: University of Chicago Press, 2005.

Gowland, Angus. *The Worlds of Renaissance Melancholy: Robert Burton in Context*. Ideas in Context 78. Cambridge: Cambridge University Press, 2006.

Grafton, Anthony. *What Was History? The Art of History in Early Modern Europe*. Cambridge: Cambridge University Press, 2007.

Grew, Nehemiah. *The Anatomy of Plants. With an Idea of a Philosophical History of Plants, and Several Other Lectures, Read before the Royal Society*. London, 1682.

———. *Cosmologia Sacra; or, A Discourse of the Universe as it is the Creature and Kingdom of God*. London, 1701.

———. *An idea of a phytological history propounded together with a continuation of the anatomy of vegetables, particularly prosecuted upon roots*. London, 1673.

———. *Musæum Regalis Societatis; or, A Catalogue & Description of the Natural and Artificial Rarities Belonging to the Royal Society*. London, 1681.

Grindle, Nick. "'No Other Sign or Note than the Very Order': Francis Willughby, John Ray and the Importance of Collecting Pictures." *Journal of the History of Collections* 17, no. 1 (2005): 15–22.

Grote, Simon. *The Emergence of Modern Aesthetic Theory: Religion and Morality in Enlightenment Germany and Scotland*. Ideas in Context 117. Cambridge: Cambridge University Press, 2017.

Guerlac, Henry, and Margaret C. Jacob. "Bentley, Newton, and Providence: The Boyle Lectures Once More." *Journal of the History of Ideas* 30, no. 3 (1969): 307–18.

Guerrini, Anita. *Obesity and Depression in the Enlightenment: The Life and Times of George Cheyne*. Norman: University of Oklahoma Press, 2000.

Guyer, Paul. *A History of Modern Aesthetics*. 3 vols. Cambridge: Cambridge University Press, 2014.

Hall, Bryan. “Kant on Newton, Genius, and Scientific Discovery.” *Intellectual History Review* 24, no. 4 (2014): 539–56.

Hamon, Philippe, and Patricia Baudoin. “Rhetorical Status of the Descriptive.” In “Towards a Theory of Description.” Special issue, *Yale French Studies* 61 (1981): 1–26.

Hanson, Craig Ashley. *The English Virtuoso: Art, Medicine, and Antiquarianism in the Age of Empiricism*. Chicago: University of Chicago Press, 2009.

Harrison, Peter. *The Fall of Man and the Foundations of Science*. Oxford: Oxford University Press, 2007.

———. “Newtonian Science, Miracles, and the Laws of Nature.” *Journal of the History of Ideas* 56, no. 4 (1995): 531–53.

———. “Physico-Theology and the Mixed Sciences: The Role of Theology in Early Modern Natural Philosophy.” In *The Science of Nature in the Seventeenth Century: Patterns of Change in Early Modern Natural Philosophy*, edited by Peter R. Anstey and John A. Schuster, 165–83. Studies in History and Philosophy of Science 19. Dordrecht, Neth.: Springer, 2005.

———. *The Territories of Science and Religion*. Chicago: University of Chicago Press, 2015.

———. “Voluntarism and Early Modern Science.” *History of Science* 40, no. 1 (2002): 63–89.

Harwood, John T. “Rhetoric and Graphics in *Micrographia*.” In *Robert Hooke: New Studies*, edited by Michael Hunter and Simon Schaffer, 119–47. Woodbridge, UK: Boydell Press, 1989.

———. “Science Writing and Rhetorical Science: Boyle and Rhetorical Theory.” In Michael Hunter, *Robert Boyle Reconsidered*, 37–56.

Hattab, Helen. *Descartes on Forms and Mechanisms*. Cambridge: Cambridge University Press, 2009.

Henry, John. “Metaphysics and the Origins of Modern Science: Descartes and the Importance of Laws of Nature.” *Early Science and Medicine* 9, no. 2 (2004): 73–114.

———. “Voluntarist Theology at the Origins of Modern Science: A Response to Peter Harrison.” *History of Science* 47, no. 1 (2009): 79–113.

Hermann, Paul. *Paradisus Batavus, Continens Plus Centum Plantas*. Leiden, Neth., 1698.

Heyd, Michael. *“Be Sober and Reasonable”: The Critique of Enthusiasm in the Seventeenth and Early Eighteenth Centuries*. Leiden, Neth.: Brill, 1995.

Hobbes, Thomas. *Leviathan*. Edited by Richard Tuck. Revised student ed. Cambridge Texts in the History of Political Thought. Cambridge: Cambridge University Press, 1996.

Hobbes, Thomas, and Bernard Lamy. *The Rhetorics of Thomas Hobbes and Bernard Lamy*. Edited and with an introduction by John T. Harwood. Carbondale: Southern Illinois University Press, 1986.

Hoffmann, Viktoria von. *From Gluttony to Enlightenment*. Urbana: University of Illinois Press, 2016.

Hooke, Robert. “Figures Observ'd in Snow by Mr. Hook.” Royal Society Register Book, vol. 2, p. 62. Royal Society Centre for the History of Science.

———. “A General Scheme, or Idea of the Present State of Natural Philosophy, and how its Defects may be Remedied by a Methodical Proceeding in the making Experiments and collecting Observations. Whereby to Compile a Natural History, as the Solid Basis for the Superstructure of True Philosophy.” In Hooke, *Posthumous Works*, 3–70. London, 1705.

———. "Lectures of Light, Explicating Its Nature, Properties, and Effects." In Hooke, *Posthumous Works*, 71–148.

———. *Micrographia; or, Some Physiological Descriptions of Minute Bodies*. London, 1665.

———. "Observables in frozen Urine." Royal Society Register Book, vol. 2, p. 61.

———. "Observables in the six-branch'd Figures in frozen Urine. By Mr. Hook, Decembr. the 10th 1662." Royal Society Register Book, vol. 2, pp. 59–61.

———. *The Posthumous Works of Robert Hooke*. Edited by Richard Waller. London, 1705.

Horace. *Satires. Epistles. The Art of Poetry*. Translated by H. Rushton Fairclough. Loeb Classical Library 194. Cambridge, MA: Harvard University Press, 1926.

Hunt, William. *A Mathematical Companion; or, The Description and Use of a New Sliding-Rule*. London, 1697.

Hunter, Matthew C. "Experiment, Theory, Representation: Robert Hooke's Material Models." In *Beyond Mimesis and Convention: Representation in Art and Science*, edited by Roman Frigg and Matthew C. Hunter, 193–219. Boston Studies in the Philosophy of Science 262. Dordrecht, Neth.: Springer, 2010.

———. "Robert Hooke Fecit: Making and Knowing in Restoration London." PhD diss., University of Chicago, 2007.

———. *Wicked Intelligence: Visual Art and the Science of Experiment in Restoration London*. Chicago: University of Chicago Press, 2013.

Hunter, Michael. "Early Problems in Professionalizing Scientific Research: Nehemiah Grew (1641–1712) and the Royal Society, with an Unpublished Letter to Henry Oldenburg." *Notes and Records of the Royal Society of London* 36, no. 2 (1982): 189–209.

———, ed. *Robert Boyle: By Himself and His Friends*. London: William Pickering, 1994.

———, ed. *Robert Boyle Reconsidered*. Cambridge: Cambridge University Press, 1994.

———. *Robert Boyle, 1627–91: Scrupulosity and Science*. Woodbridge, UK: Boydell Press, 2000.

———. "The Royal Society and the Origins of British Archaeology: I." *Antiquity* 45, no. 178 (1971): 113–21.

———. "The Royal Society and the Origins of British Archaeology: II." *Antiquity* 45, no. 179 (1971): 187–92.

Hurlbutt, Robert H. III. *Hume, Newton, and the Design Argument*. Lincoln: University of Nebraska Press, 1965.

Hutcheson, Francis. *An Inquiry into the Original of Our Ideas of Beauty and Virtue*. London, 1725.

Iliffe, Rob. "Isaac Newton: Lucatello Professor of Mathematics." In Lawrence and Shapin, *Science Incarnate*, 121–55.

Jardine, Lisa. "Hooke the Man: His Diary and His Health." In *London's Leonardo: The Life and Work of Robert Hooke*, edited by Jim Bennett, Michael Cooper, Michael Hunter, and Lisa Jardine, 163–206. Oxford: Oxford University Press, 2003.

Johns, Adrian. *The Nature of the Book: Print and Knowledge in the Making*. Chicago: University of Chicago Press, 1998.

Johnston, John. *Historiae Naturalis de Piscibus et Cetis Libri V*. Amsterdam, 1657.

Jones, Inigo, and John Webb. *The most notable Antiquity of Great Britain vulgarly called* STONE-HENG *on Salisbury Plain. Restored by* INIGO JONES *Esquire, Architect Generall to the late* KING. London, 1655.

Jones, Matthew L. *The Good Life in the Scientific Revolution: Descartes, Pascal, Leibniz, and the Cultivation of Virtue*. Chicago: University of Chicago Press, 2006.

Jones, Richard Foster. "Science and English Prose Style in the Third Quarter of the Seventeenth Century." *PMLA* 45, no. 4 (1930): 977–1009.

Kant, Immanuel. *Critique of the Power of Judgment*. Edited by Paul Guyer. Translated by Paul Guyer and Eric Matthews. Cambridge: Cambridge University Press, 2000.

Kivy, Peter. *The Seventh Sense: Francis Hutcheson and Eighteenth-Century British Aesthetics*. 2nd ed. Oxford: Oxford University Press, 2003.

Klein, Lawrence E. *Shaftesbury and the Culture of Politeness: Moral Discourse and Cultural Politics in Early Eighteenth-Century England*. Cambridge: Cambridge University Press, 1994.

Koyré, Alexandre. *Etudes Galiléennes*. Paris: Hermann, 1939.

Kusukawa, Sachiko. "The *Historia Piscium* (1686)." *Notes and Records of the Royal Society of London* 54, no. 2 (2000): 179–97.

———. *Picturing the Book of Nature: Image, Text, and Argument in Sixteenth-Century Human Anatomy and Medical Botany*. Chicago: University of Chicago Press, 2012.

Laborie, Lionel. *Enlightening Enthusiasm: Prophecy and Religious Experience in Early Eighteenth-Century England*. Manchester, UK: Manchester University Press, 2015.

Lamy, Bernard. *The Art of Speaking*. London, 1676.

———. *De l'art de parler*. Paris, 1675.

Lawrence, Christopher, and Steven Shapin, eds. *Science Incarnate: Historical Embodiments of Natural Knowledge*. Chicago: University of Chicago Press, 1998.

Lee, Rensselaer W. "*Ut Pictura Poesis*: The Humanistic Theory of Painting." *Art Bulletin* 22, no. 4 (1940): 197–269.

Leibniz, Gottfried. *Sämtliche Schriften und Briefe*. 6th ser. Vol. 3. Darmstadt: Deutsche Akademie der Wissenschaften, 1923.

Lessing, Gotthold Ephraim. "Laocoön: An Essay on the Limits of Painting and Poetry (1766)." In *Art in Theory, 1648–1815: An Anthology of Changing Ideas*, edited by Charles Harrison, Paul Wood, and Jason Gaiger, 477–86. Oxford: Blackwell, 2000.

Levine, Joseph M. *Dr. Woodward's Shield: History, Science, and Satire in Augustan England*. Ithaca, NY: Cornell University Press, 1977.

Levitin, Dmitri. "Rethinking English Physico-Theology: Samuel Parker's *Tentamina de Deo* (1665)." *Early Science and Medicine* 19, no. 1 (2014): 28–75.

Lewis, Rhodri. *Language, Mind and Nature: Artificial Languages in England from Bacon to Locke*. Cambridge: Cambridge University Press, 2007.

———. *William Petty on the Order of Nature: An Unpublished Manuscript Treatise*. Medieval and Renaissance Texts and Studies 399. Tempe: Arizona Center for Medieval and Renaissance Studies, 2012.

Lister, Martin. *The Correspondence of Dr. Martin Lister (1639–1712): Volume One, 1662–1677*. Edited by Anna Marie Roos. Medieval and Early Modern Philosophy and Science 24. Leiden, Neth.: Brill, 2015.

———. *Historiae sive Synopsis Methodicae Conchyliorum*. 5 vols. London, 1685–92.

———. "A Letter of Mr. Martyn Lister, Written to the Publisher from York, Januar. 10. 1671/2, Containing an Ingenious Account of Veins by Him Observ'd in Plants, Analogous to Human Veins." *Philosophical Transactions of the Royal Society* 6, no. 79 (1671/2): 3052–55.

Locke, John. *An Essay Concerning Human Understanding*. Edited by Peter H. Nidditch. Oxford: Oxford University Press, 1975.

Loesberg, Jonathan. "Kant, Hume, Darwin, and Design: Why Intelligent Design Wasn't Science before Darwin and Still Isn't." *Philosophical Forum* 38, no. 2 (2007): 95–123.

Louw, Hentie. "The 'Mechanick Artist' in Late Seventeenth Century English and French Architecture: The Work of Robert Hooke, Christopher Wren and Claude Perrault Compared as Products of an Interactive Science/Architecture Relationship." In *Robert Hooke: Tercentennial Studies*, edited by Michael Cooper and Michael Hunter, 181–99. Aldershot, UK: Ashgate, 2006.

Lucian of Samosata. *The Works of Lucian of Samosata*. Translated by H. W. Fowler and F. G. Fowler. 4 vols. Oxford: Clarendon Press, 1905.

Lund, Roger D. "Wit, Judgment, and the Misprisions of Similitude." *Journal of the History of Ideas* 65, no. 1 (2004): 53–75.

Lynch, William T. *Solomon's Child: Method in the Early Royal Society of London*. Stanford, CA: Stanford University Press, 2001.

MacIntosh, J. J. "Perception and Imagination in Descartes, Boyle and Hooke." *Canadian Journal of Philosophy* 13, no. 3 (1983): 327–52.

———. "Primary and Secondary Qualities." *Studia Leibnitiana* 8, no. 1 (1976): 88–104.

Mandelbrote, Scott. "Early Modern Natural Theologies." In *The Oxford Handbook of Natural Theology*, edited by John Hedley Brooke, Russell Re Manning, and Fraser Watts, 75–99. Oxford: Oxford University Press, 2013.

———. "The Uses of Natural Theology in Seventeenth-Century England." *Science in Context* 20, no. 3 (2007): 451–80.

Markley, Robert. *Fallen Languages: Crises of Representation in Newtonian England*. Ithaca, NY: Cornell University Press, 1993.

Marr, Alex. "Knowing Images." *Renaissance Quarterly* 69, no. 3 (2016): 1000–1013.

McGuinness, David. "Edward Lhuyd's Contribution to the Study of Irish Megalithic Tombs." *Journal of the Royal Society of Antiquaries of Ireland* 12, no. 126 (1996): 62–85.

Meillassoux, Quentin. *After Finitude: An Essay on the Necessity of Contingency*. Translated by Ray Brassier. London: Bloomsbury, 2012. First published 2006 as *Après la finitude* by Éditions du Seuil (Paris).

Meli, Domenico Bertoloni. *Mechanism, Experiment, Disease: Marcello Malpighi and Seventeenth-Century Anatomy*. Baltimore: Johns Hopkins University Press, 2011.

Miller, Peter N. "Description Terminable and Interminable: Looking at the Past, Nature, and Peoples in Peiresc's Archive." In Pomata and Siraisi, *Historia*, 355–97.

More, Henry. *An Antidote against Atheisme; or, An Appeal to the Natural Faculties of the Minde of Man, whether there be not a God*. London, 1653.

Mortensen, Preben. "Shaftesbury and the Morality of Art Appreciation." *Journal of the History of Ideas* 55, no. 4 (1994): 631–50.

Mullan, John. "Hypochondria and Hysteria: Sensibility and the Physicians." *Eighteenth Century* 25, no. 2 (1984): 141–74.

Mulligan, Lotte. "Robert Boyle, 'Right Reason,' and the Meaning of Metaphor." *Journal of the History of Ideas* 55, no. 2 (1994): 235–57.

———. "Self-Scrutiny and the Study of Nature: Robert Hooke's Diary as Natural History." *Journal of British Studies* 35, no. 3 (1996): 311–42.

Muri, Alison. "Enlightenment Cybernetics: Communications and Control in the Man-Machine." *Eighteenth Century* 49, no. 2 (2008): 140–63.

Newman, William R. *Atoms and Alchemy: Chymistry and the Experimental Origins of the Scientific Revolution*. Chicago: University of Chicago Press, 2006.

Nicholson, Marjorie H. *Mountain Gloom and Mountain Glory: The Development of the Aesthetics of the Infinite*. 1959. Reprint, Seattle: University of Washington Press, 1997.

Nuovo, Victor. *John Locke: The Philosopher as Christian Virtuoso*. Oxford: Oxford University Press, 2017.

Ogilvie, Brian W. "Insects in John Ray's Natural History and Natural Theology." In *Zoology in Early Modern Culture: Intersections of Science, Theology, Philology, and Political and Religious Education*, edited by Karl A. E. Enenkel and Paul J. Smith, 235–62. Intersections: Interdisciplinary Studies in Early Modern Culture 32. Boston: Brill, 2014.

———. "Natural History, Ethics, and Physico-Theology." In Pomata and Siraisi, *Historia*, 75–105.

———. *The Science of Describing: Natural History in Renaissance Europe*. Chicago: University of Chicago Press, 2006.

Oldenburg, Henry. *The Correspondence of Henry Oldenburg*. Edited and translated by A. Rupert Hall and Marie Boas Hall. 11 vols. Madison: University of Wisconsin Press, 1965–86.

Osler, Margaret J. *Divine Will and the Mechanical Philosophy: Gassendi and Descartes on Contingency and Necessity in the Created World*. Cambridge: Cambridge University Press, 1994.

———. "From Immanent Natures to Nature as Artifice: The Reinterpretation of Final Causes in Seventeenth-Century Natural Philosophy." *Monist* 79, no. 3 (1996): 388–407.

———. "The Intellectual Sources of Robert Boyle's Philosophy of Nature: Gassendi's Voluntarism and Boyle's Physico-Theological Project." In *Philosophy, Science, and Religion in England, 1640–1700*, edited by Richard W. F. Kroll, Richard Ashcraft, and Perez Zagorin, 178–98. Cambridge: Cambridge University Press, 1992.

———. "Mixing Metaphors: Science and Religion or Natural Philosophy and Theology in Early Modern Europe." *History of Science* 35, no. 1 (1997): 91–113.

———. "Whose Ends? Teleology in Early Modern Natural Philosophy." *Osiris* 16 (2001): 151–68.

Paster, Gail Kern, Katherine Rowe, and Mary Floyd-Wilson, eds. *Reading the Early Modern Passions: Essays in the Cultural History of Emotion*. Philadelphia: University of Pennsylvania Press, 2005.

Patrides, C. A. *The Cambridge Platonists*. 2nd ed. Cambridge: Cambridge University Press, 1980.

Pender, Stephen, and Nancy S. Struever, eds. *Rhetoric and Medicine in Early Modern Europe*. London: Routledge, 2012.

Pérez-Gómez, Alberto, and Louise Pelletier. *Architectural Representation and the Perspective Hinge*. Cambridge, MA: MIT Press, 1997.

[Perrault, Claude.] *Mémoires pour servir à l'histoire naturelle des animaux*. 2 vols. Paris, 1671–76.

Perrault, Claude. *Memoir's for a natural history of animals containing the anatomical descriptions of several creatures dissected by the Royal Academy of Sciences at Paris*. Translated by Alexander Pitfield. London, 1688.

Picciotto, Joanna. *Labors of Innocence in Early Modern England*. Cambridge, MA: Harvard University Press, 2010.

———. "Reforming the Garden: The Experimentalist Eden and *Paradise Lost*." *ELH* 72, no. 1 (2005): 23–78.

Piggott, Stuart. *Antiquity Depicted: Aspects of Archaeological Illustration*. London: Thames and Hudson, 1978.

Plukenet, Leonard. *Phytographia, sive Stirpium Illustriorum, et Minus Cognitarium Icones*. 6 vols. London, 1691–1705.

Plutarch. *Moralia, Volume IV: Roman Questions. Greek Questions. Greek and Roman Parallel Stories. On the Fortune of the Romans. On the Fortune or the Virtue of Alexander. Were the Athenians More Famous in War or in Wisdom?* Translated by Frank Cole Babbitt. Loeb Classical Library 305. Cambridge, MA: Harvard University Press, 1936.

Pomata, Gianna, and Nancy G. Siraisi, eds. *Historia: Empiricism and Erudition in Early Modern Europe*. Cambridge, MA: MIT Press, 2005.

Poole, William. "The Divine and the Grammarian: Theological Disputes in the 17th-Century Universal Language Movement." *Historiographica Linguistica* 30, no. 3 (2008): 273–300.

———. "Francis Lodwick's Creation: Theology and Natural Philosophy in the Early Royal Society." *Journal of the History of Ideas* 66, no. 2 (2005): 245–63.

———. *The World Makers: Scientists of the Restoration and the Search for the Origins of the Earth*. Oxford: Peter Lang, 2010.

Porter, Roy. "Consumption: Disease of the Consumer Society." In *Consumption and the World of Goods*, edited by John Brewer and Roy Porter, 58–81. London: Routledge, 1993.

Preston, Claire. *The Poetics of Scientific Investigation in Seventeenth-Century England*. Oxford: Oxford University Press, 2016.

Principe, Lawrence M. *The Aspiring Adept: Robert Boyle and His Alchemical Quest*. Princeton, NJ: Princeton University Press, 1998.

———. "Virtuous Romance and Romantic Virtuoso: The Shaping of Robert Boyle's Literary Style." *Journal of the History of Ideas* 56, no. 3 (1995): 377–97.

Purcell, John. *A Treatise of Vapours; or, Hysterick Fits*. London, 1702.

Quintilian. *Institutio Oratoria*. Translated by H. E. Butler. 4 vols. Loeb Classical Library 124–27. Cambridge, MA: Harvard University Press, 1921.

Raven, Charles. *John Ray, Naturalist: His Life and Works*. Cambridge: Cambridge University Press, 1942.

Ray, John. *Catalogus Plantarum circa Cantabrigiam Nascentium*. Cambridge, 1660.

———. *The Correspondence of John Ray*. Edited by Edwin Lankester. London: Ray Society, 1848.

———. *De Variis Plantarum Methodis Dissertatio Brevis*. London, 1696.

———. *Further Correspondence of John Ray*. Edited by Robert W. T. Gunther. London: Ray Society, 1928.

———. *Historia Plantarum*. 3 vols. London, 1686–1704.

———. *John Ray's Cambridge Catalogue (1660)*. Translated and edited by P. H. Oswald and C. D. Preston. London: Ray Society, 2011.

———. "A Letter from That Incomparable Botanist Mr. John Ray, giving an Account of the Phytographia of Leonard Plukenet, M. D. Lately published. Lond. fol. 1691." *Philosophical Transactions of the Royal Society* 17, no. 194 (1686): 528–30.

———. *Philosophical Letters between the Late Learned Mr. Ray and several of his Ingenious Correspondents*. Edited by William Derham. London, 1718.

———. *Synopsis Methodica Stirpium Britannicarum*. 2nd ed. London, 1696.

———. *Three Physico-Theological Discourses*. London, 1693.

———. *The Wisdom of God Manifested in the Works of Creation*. London, 1691.

———. *The Wisdom of God Manifested in the Works of Creation*. 2nd ed. London, 1692.

Ray, John, and Francis Willughby. *De Historia Piscium*. Oxford, 1686.

———. *The Ornithology of Francis Willughby [. . .] In Three Books, Wherein All the* BIRDS *Hitherto Known, being reduced into a Method sutable to their Natures, are accurately described*. London, 1678.

Ricciardo, Salvatore. "Robert Boyle on God's 'Experiments': Resurrection, Immortality and Mechanical Philosophy." *Intellectual History Review* 25, no. 1 (2015): 97–113.

Richardson, Jonathan. "Hymn to God." British Library Add. MS. 10423.

———. *Two Discourses. I. An Essay On the whole Art of Criticism as it relates to Painting.* [. . .] *II. An Argument in behalf of the Science of a Connoisseur.* London, 1719.

Richardson, Samuel. *Correspondence with George Cheyne and Thomas Edwards.* Edited by David E. Shuttleton and John A. Dussinger. Cambridge Edition of the Correspondence of Samuel Richardson 2. Cambridge: Cambridge University Press, 2013.

Rind, Miles. "The Concept of Disinterestedness in Eighteenth-Century British Aesthetics." *Journal of the History of Philosophy* 40, no. 1 (2002): 67–87.

Riskin, Jessica. *Science in the Age of Sensibility: The Sentimental Empiricists of the French Enlightenment.* Chicago: University of Chicago Press, 2002.

Rivers, Isabel. "'Galen's Muscles': Wilkins, Hume, and the Educational Use of the Argument from Design." *Historical Journal* 36, no. 3 (1993): 577–97.

Roberts, Brynley F. "Lhuyd [Lhwyd; formerly Lloyd], Edward (1659/60?–1709), naturalist and philologist." *Oxford Dictionary of National Biography.* Oxford: Oxford University Press, 2004.

Robson, Mark. *The Sense of Early Modern Writing: Rhetoric, Poetics, Aesthetics.* Manchester, UK: Manchester University Press, 2006.

Roos, Anna Marie. "The Art of Science: A 'Rediscovery' of the Lister Copperplates." *Notes and Records of the Royal Society* 66, no. 1 (2012): 19–40.

———. "Nehemiah Grew (1641–1712) on the Saline Chymistry of Plants." *Ambix* 54, no. 1 (2007): 51–68.

Rossi, Paolo. *The Dark Abyss of Time.* Translated by Lydia G. Cochrane. Chicago: University of Chicago Press, 1984.

Rousseau, G. S. "Nerves, Spirits, and Fibres: Towards Defining the Origins of Sensibility." In *Studies in the Eighteenth Century III: Papers Presented at the Third David Nichol Smith Memorial Seminar 1973*, edited by R. F. Brissenden and J. C. Eade, 137–57. Canberra: Australian National University Press, 1976.

Sarafianos, Aris. "Pain, Labor, and the Sublime: Medical Gymnastics and Burke's Aesthetics." *Representations* 91, no. 1 (2005): 58–83.

Schaffer, Simon. "Newtonian Angels." In *Conversations with Angels: Essays towards a History of Spiritual Communication*, edited by Joad Raymond, 90–122. Houndmills, UK: Palgrave Macmillan, 2011.

———. "Regeneration: The Body of Natural Philosophers in Restoration England." In Lawrence and Shapin, *Science Incarnate*, 83–120.

Schneewind, J. B. "Locke's Moral Philosophy." In *The Cambridge Companion to Locke*, edited by Vere Chappell, 199–225. Cambridge: Cambridge University Press, 1994.

Seigworth, Gregory J., and Melissa Gregg. "An Inventory of Shimmers." In *The Affect Theory Reader*, edited by Melissa Gregg and Gregory J. Seigworth, 1–25. Durham, NC: Duke University Press, 2010.

Shaftesbury, Third Earl of [Anthony Ashley Cooper]. *Characteristics of Men, Manners, Opinions, Times.* Edited by Lawrence E. Klein. Cambridge Texts in the History of Philosophy. Cambridge: Cambridge University Press, 2000.

———. "A Letter Concerning Enthusiasm to my Lord *****." In Shaftesbury, *Characteristics of Men, Manners, Opinions, Times*, 4–28.

———."Miscellaneous Reflections on the Preceding Treatises and Other Critical Subjects." In Shaftesbury, *Characteristics of Men, Manners, Opinions, Times*, 339–483.

Shagan, Ethan. *Popular Politics and the English Reformation*. Cambridge Studies in Early Modern British History. Cambridge: Cambridge University Press, 2003.

Shanahan, Timothy. "God and Nature in the Thought of Robert Boyle." *Journal of the History of Philosophy* 26, no. 4 (1988): 547–69.

———. "Teleological Reasoning in Boyle's *Disquisition about Final Causes*." In Michael Hunter, *Robert Boyle Reconsidered*, 177–92.

Shapin, Steven. "Descartes the Doctor: Rationalism and Its Therapies." *British Journal for the History of Science* 33, no. 2 (2000): 131–54.

———. "Pump and Circumstance: Robert Boyle's Literary Technology." *Social Studies of Science* 14, no. 4 (1984): 481–520.

———. "'A Scholar and a Gentleman': The Problematic Identity of the Scientific Practitioner in Early Modern England." *History of Science* 29, no. 3 (1991): 279–327.

———. "The Sciences of Subjectivity." *Social Studies of Science* 42, no. 2 (2011): 170–84.

———. "A Taste of Science: Making the Subjective Objective in the California Wine World." *Social Studies of Science* 46, no. 3 (2016): 436–60.

Shapin, Steven, and Simon Schaffer. *Leviathan and the Air-Pump: Hobbes, Boyle, and the Experimental Life*. Princeton, NJ: Princeton University Press, 1985.

Sheehan, Jonathan. "Thomas Hobbes, D.D.: Theology, Orthodoxy, and History." *Journal of Modern History* 88, no. 2 (2016): 249–74.

Sheehan, Jonathan, and Dror Wahrman. *Invisible Hands: Self-Organization and the Eighteenth Century*. Chicago: University of Chicago Press, 2015.

Skinner, Quentin. *Reason and Rhetoric in the Philosophy of Hobbes*. Cambridge: Cambridge University Press, 1996.

Slaughter, Mary. *Universal Languages and Scientific Taxonomy*. Cambridge: Cambridge University Press, 1982.

Sloan, Phillip R. "John Locke, John Ray, and the Problem of the Natural System." *Journal of the History of Biology* 5, no. 1 (1972): 1–53.

Smith, Courtney Weiss. *Empiricist Devotions: Science, Religion, and Poetry in Early Eighteenth-Century England*. Charlottesville: University of Virginia Press, 2016.

———. "Rhyme and Reason in John Wilkins's Philosophical Language Scheme." *Modern Philology* 115, no. 2 (2017): 183–212.

Smith, Jeffrey Chipps, ed. *Visual Acuity and the Arts of Communication in Early Modern Germany*. Aldershot, UK: Ashgate, 2014.

Spary, Emma. *Eating the Enlightenment: Food and the Sciences in Paris, 1670–1760*. Chicago: University of Chicago Press, 2012.

Spinoza, Baruch. *Ethics*. Edited and translated by G. H. R. Parkinson. Oxford Philosophical Texts. Oxford: Oxford University Press, 2000.

———. *Opera Posthuma*. Amsterdam, 1677.

Sprat, Thomas. *The History of the Royal-Society of London for the Improving of Natural Knowledge*. London, 1667.

———. *Observations into Mons. De Sorbiere's Voyage into England*. London, 1665.

Stafford, Barbara Maria. *Artful Science: Enlightenment Entertainment and the Eclipse of Visual Education*. Cambridge, MA: MIT Press, 1994.

Stalnaker, Joanna. *The Unfinished Enlightenment: Description in the Age of the Encyclopedia*. Ithaca, NY: Cornell University Press, 2010.

Steno, Nicolas. *Discours de Monsieur Stenon, sur L'Anatomie du Cerveau*. Paris, 1669.

Stolnitz, Jerome. "On the Origins of 'Aesthetic Disinterestedness.'" *Journal of Aesthetics and Art Criticism* 20, no. 2 (1961): 131–43.

Swan, Claudia. "The Uses of Realism in Early Modern Illustrated Botany." In *Visualising Medieval Medicine and Natural History*, edited by Jean Givens, Karen M. Reeds, and Alain Touwaide, 239–50. Aldershot, UK: Ashgate, 2006.

Sydenham, Thomas. *The Compleat Method of Curing Almost all Diseases*. London, 1694.

———. *Processus Integri in Morbis fere Omnibus Curandis*. London, 1692.

Townsend, Dabney. "From Shaftesbury to Kant: The Development of the Concept of Aesthetic Experience." *Journal of the History of Ideas* 48, no. 2 (1986): 287–305.

Tuan, Yi-Fu. *The Hydrologic Cycle and the Wisdom of God: A Theme in Geoteleology*. Toronto: University of Toronto Press, 1968.

Vickers, Brian. "The Royal Society and English Prose Style: A Reassessment." In *Rhetoric and the Pursuit of Truth: Language Change in the Seventeenth and Eighteenth Centuries*, edited by Brian Vickers and Nancy S. Struever, 1–76. Los Angeles: Clark Memorial Library, University of California, 1985.

Vidal, Fernando. "Brains, Bodies, Selves, and Science: Anthropologies of Identity and the Resurrection of the Body." *Critical Inquiry* 28, no. 4 (2002): 930–74.

Vidal, Fernando, and Bernhard Kleeberg. "Introduction: Knowledge, Belief, and the Impulse to Natural Theology." *Science in Context* 20, no. 3 (2007): 381–400.

Waddell, Mark A. *Jesuit Science and the End of Nature's Secrets*. London: Routledge, 2015.

Walker, Matthew. *Architects and Intellectual Culture in Post-Restoration England*. Oxford: Oxford University Press, 2017.

Walmsley, Peter. *Locke's Essay and the Rhetoric of Science*. Lewisburg, PA: Bucknell University Press, 2003.

Wear, Andrew. "The Spleen in Renaissance Anatomy." *Medical History* 21, no. 1 (1977): 43–60.

Webb, John. *A Vindication of Stone Heng Restored*. 2nd ed. London, 1725.

Westfall, Richard S. *Science and Religion in Seventeenth-Century England*. 1958. Reprint, Ann Arbor: University of Michigan Press, 1973.

Wilkins, John. *Of the Principles and Duties of Natural Religion*. London, 1678.

Williams, Kelsey Jackson. *The Antiquary: John Aubrey's Historical Scholarship*. Oxford: Oxford University Press, 2016.

Willis, Thomas. *Cerebri Anatome, cui accessit Nervorum Descriptio et Usus*. London, 1664.

———. *De Anima Brutorum*. Oxford, 1672.

———. *Dr. Willis's Practice of Physick, Being the whole Works of that Renowned and Famous Physician*. Translated by Samuel Pordage. London, 1684.

Wilson, Catherine. *Epicureanism at the Origins of Modernity*. Oxford: Oxford University Press, 2017.

———. *Leibniz's Metaphysics: A Historical and Comparative Study*. Princeton, NJ: Princeton University Press, 2015.

Wojcik, Jan W. *Robert Boyle and the Limits of Reason*. Cambridge: Cambridge University Press, 1997.

Wolfe, Charles T., and Ofer Gal, eds. *The Body as Object and Instrument of Knowledge: Embodied Empiricism in Early Modern Science*. Studies in History and Philosophy of Science 25. Dordrecht, Neth.: Springer, 2010.

———. "Embodied Empiricism." In Wolfe and Gal, *Body as Object*, 1–5.

Wolfe, Charles T., and Michaela van Esveld. "The Material Soul: Strategies for Naturalising the

Soul in an Early Modern Epicurean Context." In *Conjunctions of Mind, Soul and Body from Plato to the Enlightenment*, edited by Danijela Kambaskovic, 371–421. Dordrecht, Neth.: Springer, 2014.

Wragge-Morley, Alexander. "A Strange and Surprising Debate: Mountains, Original Sin and 'Science' in Seventeenth-Century England." *Endeavour* 33, no. 2 (2009): 76–80.

Wren, Christopher. *Wren's "Tracts" on Architecture and Other Writings*. Edited and with an introduction by Lydia M. Soo. Cambridge: Cambridge University Press, 1998.

Yolton, John W. *Locke and the Compass of Human Understanding: A Selective Commentary on the "Essay."* Cambridge: Cambridge University Press, 1970.

Zeitz, Lisa M. "Addison's 'Imagination' Papers and the Design Argument." *English Studies* 73, no. 6 (1992): 493–502.

Zitin, Abigail. "Thinking Like an Artist: Hogarth, Diderot, and the Aesthetics of Technique." *Eighteenth-Century Studies* 46, no. 4 (2013): 555–70.

Index

Page numbers in italics refer to illustrations.